A-Z
BOTANY

A-Z BOTANY

Prof. Gyan Deep Singh

CENTRUM PRESS
NEW DELHI-110002 (INDIA)

CENTRUM PRESS
H.O.: 4360/4, Ansari Road, Daryaganj,
New Delhi-110 002 (India)
Ph.: 23278000, 23261597

B.O.: No. 1015, Ist Main Road, BSK IIIrd Stage
IIIrd Phase, IIIrd Block,
Bangalore - 560 085 (India)
Tel.: 080-41723429
Visit us at: www.centrumpress.com

A-Z Botany

First Edition, 2009
ISBN 978-93-80106-09-0

PRINTED IN INDIA

Printed at Salasar Imaging Systems, Delhi-110035 (India)

Contents

Preface

Botany is a branch of biology and is the scientific study of plant life and development. Botany covers a wide range of scientific disciplines that study plants, algae, and fungi including: structure, growth, reproduction, metabolism, development, diseases, and chemical properties and evolutionary relationships between the different groups. Botany, the study of plants, began with tribal efforts to identify edible, medicinal and poisonous plants, making botany one of the oldest sciences.

From this ancient interest in plants, the scope of botany has increased to include the study of over 550,000 kinds or species of living organisms. A considerable amount of new knowledge today is being generated from studying model plants like *Arabidopsis thaliana*. This weedy species in the mustard family was one of the first plants to have its genome sequenced. The sequencing of the rice genome and a large international research community have made rice the de facto cereal/grass/monocot model.

Another grass species, *Brachypodium distachyon* is also emerging as an experimental model for understanding the genetic, cellular and molecular biology of temperate grasses. Other commercially-important staple foods like wheat, maize, barley, rye, pearl millet and soybean are also having their genomes sequenced.

Author

Chapter 1

Green Colour of Plants

Imagine for a moment that you are looking at the earth from a point somewhere between here and the moon. You see an immense bright globe hanging in the sky. As it spins you can see that most of its surface is covered with dark blue irregular expanses of water. The land, here thrown up into a complex system of folds and wrinkles, there extending for hundreds of miles in flat or gently undulating plains, is mostly green.

Thousands of square miles are hidden under a mantle of leafy trees, shrubs and vines. Over other thousands of miles tall grasses ripple in the wind. Here and there you see large areas cleared by man and replanted with greenery of his own choosing: grains, vegetables, fruits, set in orderly rows and squares.

There are regions sparsely covered with the greyish green of sagebrush and greasewood or the fantastic forms and bristling spears of agave and cactus. In some zones bare trunks and branches offer only a promise of foliage; a few expanses of drifting sands are destitute of green cover; the tops of some of the highest wrinkles are bare rock or white snow. But as a whole the land is green. The colour of the landscape is perhaps its most remarkable feature; so uniform that we do not generally remark it.

Here and there we grow flowers of splendid and varied colours and when, in temperate regions, the frosty nights and failing sun of autumn work their annual transformations, we do not fail to admire the display of reds, browns and yellows. But these are the signals of death and of new life, of the fall of the leaf and the spring of the year. The everyday colour of life is about us most of the time; we pause not to think what a world would be like coloured otherwise, perhaps pink or blue. The basic fact of our life is that the world is green and we live upon it only because it is green.

The uniformity of the world is even more surprising than its diversity. We are more or less accustomed to comment- ing on the vast and heterogeneous assemblages that are the living plants and animals of the earth; a variety that becomes the more staggering as biology labors to bring it into some sort of order. It seems the more remarkable that biology does in fact succeed in pointing out the most unexpected resemblances in the midst of diversity. The bones of a seal's flipper are arranged in much the same way as those of your hand and those of a bird's wing can be identified with the skeleton of a man's arm. In the elements of structure a rose is related to a mushroom; in details of reproductive mechanism a pine tree resembles a moss.

Almost all the green substances of the plant world are so closely related chemically that they are commonly called by one name: chlorophyll, leaf-green. By this one word we refer to the pigments of living creatures so different as oaks and the grasses at their feet; so widely separated as the flowering plants of your garden and the microscopic plants that swarm in ponds, lakes and streams. All are green in the same way and for the same reason.

Chemists distinguish two types of chlorophyll, one a bluish, the other a yellowish green. These exist in different kinds of plants in different proportions, generally associated with pigments of other colours, yellow and orange and often blue, red, or brown. The proportions of the chlorophylls and of the other pigments are characteristic of various species of plants, so that leaves of grass are of a different green from leaves of elm and bluegrass differs in tint from timothy.

The chlorophylls cannot be extracted from a leaf by soaking it in water or even by boiling it in water (though some may escape from broken parts and the green juice may be pressed out of crushed leaves so as to leave a stain on white garments). It is easy, however, to treat green leaves with alcohol or acetone or some other organic solvent so as to obtain a beautiful clear green liquid. Such an extract contains dissolved chlorophyll (somewhat changed chemically). When the extract is poured off the leaves, they may be left quite blanched.

This is the stuff, this chlorophyll, that has been called the most important material in the world. If through some inconceivable accident all the chlorophyll of the earth should disappear today and no more be formed, all the activity of animals, all the work and worrying of humanity, all art and science and buying and selling and loving and making war,—all life would cease as soon as present stores of food were exhausted. Is it fantastic to assert that there is a relation between the activity of my fingers as I write these words, of your eyes as you read them and the fact that plants con- tain a green pigment? To make this clear necessitates a digression.

The central fact of vital activity is a ceaseless consumption of food. Ceaseless even when we are not actually eating, the minute units of which we are composed are busy with the food of our last meal, distributed to them by blood. Our life depends upon food; continual supplies of new food must be acquired and delivered to the insatiable creature within. What becomes of that food and how comes it that there is an apparently unending supply in the world?

Food is used up in a living body in a manner in many ways analogous to burning and we may use this familiar change to explain the activity of a living creature. As I put a fresh piece of wood on the fire, I point out how the wood goes into fire. "No," says the little girl, "That isn't right. The fire goes into the wood, not the wood into the fire." But with further explanation she admits that after a while the wood has disappeared; it has indeed gone somewhere.

All that is left is a little heap of ashes; the smoke and other gases —some of them for a brief moment brightly shining—have disappeared up the chimney. Many children have seen, in some "science course," the lighted candle balanced with weights; as the candle burns it rises higher and higher, the weights sink lower and lower. The actual mass of the candle is transformed into smoke and gases which vanish into the surrounding air. One of the gases commonly formed by combustion is water; water exists in the air as an invisible gas as well as on the earth as a visible liquid. Another gas generally formed when something is burned is a tasteless, colourless, odorless gas called carbon dioxide. It is a universal ingredient of air and is familiar, in more concentrated form, as the bubbles in sparkling beverages.

The same substances are formed by living bodies and usually appear as waste materials. Water is given off in our breath; a frosty morning is all that is needed for experimental demonstration. It is easy also to show that carbon dioxide is exhaled. It is known that lime water becomes milky when carbon dioxide passes into it; breathing into it causes the same transformation. It can be shown also that as these materials appear and are given off from the body, the amount of food within the body diminishes proportionally.

There is an exact relationship between the quantity of food that disappears and the quantity of carbon dioxide and water that appears. Food is indeed used in the body in other ways also; some of it is used in the construction of new parts, some may accumulate as fat. But some is constantly being decomposed into comparatively simple ingredients which are eliminated from the body. We are all chemists unawares; the solid piece of toast which we eat for breakfast passes later in the day from our mouth or nostrils as a mix- ture of impalpable gases.

In the process of burning, the gas oxygen is taken from the air and used in the chemical transformation of wax or paraffin or wood into carbon dioxide and water. The same gas is used in the corresponding activity of living matter. It is common knowledge that a lack of oxygen results in a short time in the cessation of our vital activities; no better demonstration can be made of the unending necessity of the process by which food is consumed—burned up, as it were-in the body.

In this respect plants are similar to men. By enclosing some vigorously growing plants (young seedlings are convenient) in a glass jar which is tightly stoppered, we can show, by appropriate chemical and physical tests, that the oxygen within the vessel disappears in proportion as carbon dioxide and water appear and as the solid matter of the plants diminishes. It is sometimes said indeed that plants breathe. This is obviously not true; they have no lungs, no means at all of inhaling and exhaling air.

But it is true that the life-process which necessitates our breathing occurs also in plants (though considerably more slowly). To this universal process of food-consumption the biologist refers by the word respiration. Biologically this term refers to the chemical transformations within the units of living matter. (The physician uses the word in a different sense to mean the action of the lungs.) In the fact that they respire men do not differ essentially from jellyfish and blades of grass are the equals of the cattle which crop them.

Many persons speak of "life" as something possessed only by animals; "biology" to them means the study of animals; plants, if indeed alive, are not so to the same degree or in the same sense. Even Stephen Hales, pioneer in physiology, spoke of plants as "inanimate." The error of such a distinction isapparent from a knowledge of the generality of this one process, respiration.

In all the burning, the consumption, the respiration, which goes on in and around us in nature, is a very significant feature; a feature indeed intrinsic in the structure of this universe. The substances formed by burning have lost something in the process of getting themselves formed; they have lost partly or wholly the ability to burn. You can collect the gases and smoke formed by the burning of a candle; you cannot get more gases and more smoke by burning them in turn. Burning commonly produces heat and light; that is to say, it yields energy of various types.

A definite quantity of a substance will in burning release a definite quantity of energy, which is dissipated as heat and light; no more can be obtained in this way from the materials which originally composed that substance. Exactly the same is true of life. We are fortunately able to use as food a great and pleasing variety of substances, most of which we transform by our skillful inner chemistry into carbon dioxide, water and certain other wastes. The waste materials may be collected in large quantities, but are of no value as food. A bottle of "pop" indeed consists mainly of water and carbon dioxide; but, while it may allay our thirst and titillate our palates, it cannot be thought of as nutritious.

Water is necessary to us in a variety of ways; but it is not transformed into new wastes, with a further release of energy. For food we must turn again to the sources of supply—seemingly end- less—from which we have already drawn food, the plants and animals of our domestic economy and take from them the meats, the fats, the starches and sugars and the other agreeable things which will keep us going yet another day.

With such facts in mind it is possible to divide all the materials of the world into two categories: the foods and the non-foods. The foods are starches, sugars, celluloses, proteins, fats, various acids and alcohols. The non-foods include the simple gases of the air, such as oxygen, nitrogen and carbon dioxide; water and the minerals and metals of the earth. In the early days of chemistry it was beyond human skill to manufacture foods from the latter class of substances; their synthesis was considered to be a peculiarity of living "organisms," and they were accordingly named "organic." The term has lived, although the distinction can no longer be precisely drawn.

Chemists have learned to make many of the simpler foods and so have learned that there is no essential mystery in their construction. But foods and other substances that will burn are still called organic substances and the chemistry of their behaviour is organic chemistry; other substances are inorganic.

The word organic now refers to the molecular structure of the substance rather than to its origin or to its use by organisms. Organic substances all contain the element carbon (some inorganic ones do also). Organic substances mostly have large and complex ultimate particles (molecules). They generally burn, usually with the formation of carbon dioxide.

Many of them, such as petroleum and its legion products, are scarcely thought of as foods; but all foods (with a few minor exceptions among those of the most minute living things) are organic substances.

In brief, the substance of the preceding paragraphs is that life involves a continual change of organic substances into inorganic. If this were all the picture, it would be evident that the world was rather rapidly running down, a vast clock with no one to wind it. The quantity of matter in the earth is not infinite and of this limited quantity only a certain fraction is organic. This leads us back again to a consideration of the greenness of the world; chlorophyll is the key by which the clock is wound. In chlorophyllous plants and only in them, organic substances are continually being synthesized from inorganic substance, at a rate which balances their destruction by all the living bodies of the earth.

The fact that food is actually manufactured in a green leaf may be easily demonstrated by the simple experiments used in many a schoolroom. A drop of iodine solution on a starched garment makes a spot which is not brown but blue or violet. This change may be used to reveal the presence of starch. If a leaf is bleached (by means of alcohol or acetone) and treated with iodine, it changes to a deep blue in colour; it is filled with starch.

The starch can be seen in the cells of the leaf by careful study with a microscope; it exists as minute solid granules of characteristic form. This starch is not supplied from outside; there is none in the soil nor in the air around the plant; yet as the plant grows and forms new leaves, each leaf becomes successively filled with starch. Starch also appears in other parts of the plant:—in the tubers of a potato-plant, in grains of wheat, in bananas and beans; this is surplus starch, which was made first in the green leaves of the plant (or at least partly made there), then transferred to the part in which it accumulates.

Some leaves fail to turn blue when treated with iodine, even when they have grown under the conditions most favorable to the manufacture of food and were when gathered in perfect health. They are not lacking in food nor in the ability to make food. Suitable chemical tests may be devised (with substances other than iodine) to show that their product is a sugar, commonly called glucose or dextrose. Indeed, it is generally thought that this is the first food made in all leaves and that it is changed into starch in those leaves in which starch is found. Glucose seems to be the first recognizable organic product made by green leaves from inorganic materials.

Is it possible to detect the materials which are used in the manufacture of sugar in a leaf? Certainly this food cannot be made from nothing, even by the compulsion of a living plant. In the endless cycle of nature, food is made from those same materials which are formed when it is burned or used up by a living body. The food in a leaf was made from things as tenuous and unappetizing as the water of the soil and the carbon dioxide of the air.

Water, like other things, is composed of those units called atoms; in particular it is composed of atoms of hydrogen and oxygen. These atoms, incredibly small, are grouped into the minute particles known as molecules, the particles which make up the mass of a substance, which *are* the substance. If we designate hydrogen and oxygen by their chemical symbols H and O, we may describe a single molecule of water by the formula known to every schoolchild—H_2O; this means that it contains two atoms of hydrogen and one of oxygen. A molecule of carbon dioxide is CO_2.

By the chemical transformations accomplished within the green leaf, the same atoms are juggled into molecules each of which may be represented by the more imposing formula $C_6H_{12}O_6$; this group of 24 atoms is a molecule of sugar (glucose). When chemical substances (and all substances are chemical) are changed into new substances, the atoms are not essentially changed, nor are any lost or created; they are separated and reassembled into new combinations. This is what happens when things burn, when foods are consumed in living bodies or built into new parts of living bodies. This is what happens in the manufacture of food by the green leaf. In this action any six molecules of carbon dioxide can combine with any six of water to form a single molecule of sugar; organic material is formed in the leaf at exactly the same rate as the corresponding amounts of water and carbon dioxide are used up.

The proportions involved may be readily understood if we write out the process chemist-wise as an equation, in which the same atoms are seen in the old combinations and in the new:

$$6CO_2 + 6H_2O = C_6H_{12}O_6 + 6O_2$$

This is not a description of what actually happens. Probably the manufacture of sugar ($C_6H_{12}O_6$) occurs in several steps, the products of one chemical reaction being the raw materials of the next; but the final result is that depicted by the equation and the proportions of the original materials and of the final product are as there shown.

The equation shows that besides sugar oxygen (O_2) is manufactured. It is easy to show that plants actually make gaseous oxygen. Observe a water plant growing in a glass jar exposed to the sunlight. Bubbles quickly become visible on the surface of the plant; they become detached and rise through the water and fresh bubbles take their place. The bubbles may easily be collected in a closed vessel and by a simple chemical trick proved to consist of oxygen. This is the gas so necessary to our own respiration and continually removed from the atmosphere by our activity. The same process of living plants provides us not only with the food we eat but with the essential constituent of the air we breathe.

The further change of sugar into starch which occurs commonly in most green leaves takes place through a slight change in the proportions of the atoms in the molecules of sugar. If from each molecule of glucose ($C_6H_{12}O_6$) sufficient atoms of hydrogen (H) and oxygen (O) are removed to form a molecule of water (H_2O), a residue is formed which has the chemical formula $C_6H_{10}O_5$. This does not itself become a single molecule; but a number of such atom-blocks unite to compose an enormous and complex structure which is a molecule of starch. In this the proportions of the three kinds of atoms are the same—6:10:5. The total number of such sugar-residues involved in the formation of one molecule of starch is extremely difficult to ascertain with precision; there are different kinds of starch, in which the number may differ. It is probably more than 35; a molecule of starch may consist of more than 700 atoms!

It is necessary now to reconsider the statement made above that a green leaf, properly treated with iodine, will demonstrate by a striking change in colour that it contains starch. This assertion must be qualified in various ways. The leaf must be vigorous and healthy. It must be exposed to the air from which it draws its supply of carbon dioxide. It must be well supplied, through its stem and roots, with water. Furthermore, it is a fact that the test will not work well if it is made early in the morning.

During the night the starch made in the preceding day disappears from the leaf, some being used up in the life-activity of the leaf, the rest being transported to other parts of the plant. And during the night also no new starch—nor sugar nor oxygen-is made. It may be shown, in fact, that the process by which food is manufactured, this synthesis upon which we all depend for our bread and butter, is absolutely dependent upon light; it is a Photosynthesis, a manufacture-by-light.

If a portion of a living leaf is screened from light (but otherwise kept normal and healthy) and if the entire leaf is exposed to the sun for a few hours, the screened part fails to colour blue when treated with iodine, while the surrounding unscreened parts behave as usual.

Light is not one of the materials from which food is made. Indeed, as currently understood, light is not a material at all; it is energy. It cannot be arrested in its flight and held motionless, it cannot be accumulated and stored in a paper sack or a metal cylinder. The materials of photosynthesis, the actual atoms used, have already been completely accounted for; they are carbon, hydrogen and oxygen, obtained combined as carbon dioxide and water. Light is the energy which brings about the rearrangement of the atoms into two new substances.

The force of contraction in the muscles of a man arranges bricks and beams into a wall; a stream of falling water turns the wheels of a mill; the expansion of steam drives pistons and moves tons of steel; an electric current causes the movement of the clapper of a bell; so do the moving waves of light effect a chemical transformation in green leaves. If it seems strange that something so impalpable as light can move atoms and make actual ponderable substances, think for a moment of the nature of the energy involved in any ordinary work. What is it that actually passes along a thin wire and causes the movement of a metal rod?

What indeed is the terrific force exerted by a molecule of water heated to the boiling point, or by the expanding gases of exploding dynamite? What indeed is the actual force which you impart to a body when you push it into a place where you want it? You can *feel* the action of your muscles, you are conscious of the force which the body exerts upon your hands; but do you actually transfer anything, any part of yourself, to the moved object? What passes from you to it? To answer such questions (even partially) necessitates a study of physics.

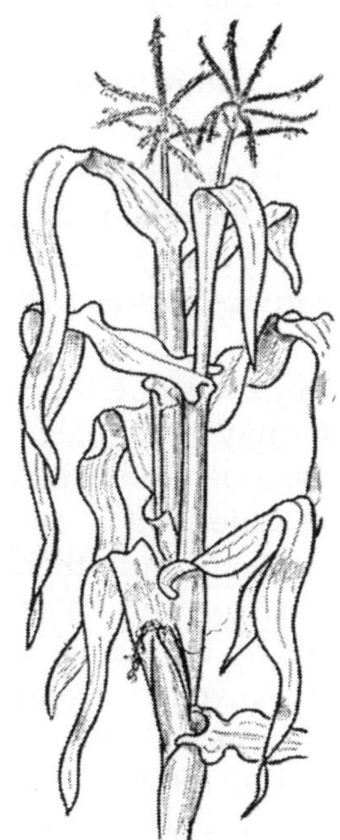

Fig. Long Leaf of Maize Plant

The distinction between matter and energy is not one which can be phrased in descriptive terms intelligible to the layman and even the physicists sometimes quarrel about these concepts. But for ordinary purposes we may recognize that the universe may be described in terms of particles, which account for its structure and of energy, which accounts for its activity. What is inseparable in nature we may separate in science. So a photographic film contains and is composed of certain atoms in certain combinations.

Light impinging upon the film causes a rearrangement of the atoms so that new substances are formed. In the same way certain atoms exist in a leaf, combined in inorganic molecules. Under the influence of light and not in its absence, that rearrangement, that chemical change occurs which results in the manufacture from these materials of the organic substances called foods and of the gaseous oxygen which we breathe.

One of the unexpected attributes of nature is her numerical propensity. Even energy may be measured, in terms of the amount of work done and we find that the appearance or disappearance of given quantities of chemical energy is correlated with changes in a precise number of molecules. A leaf (in which photosynthesis occurs) is capable of receiving a given amount of energy from the sunlight that falls upon it; the larger its area the more energy it can absorb. A potato leaf with an area of five square inches makes about a gram of glucose in a month.

A man may use, in the same time, the sugar made by 30,000 such leaves. He may not, indeed, eat so much sugar; but all the food in his potatoes and in all the other comestibles which come to his table is derived from the sugar made in the leaves of plants, potatoplants and others. A definite number of leaves, whether grown in the fields or in a glasshouse, whether nourished directly by the soil or by the ingredients of soil purified and dissolved in water to make a nutrient solution,—a definite area of leaf surface is necessary for the support of one man for one month.

As might be expected light, which varies markedly in intensity, varies also in the rate at which it will cause sugar to be made. Photosynthesis is at a maximum during the intense light of a cloudless day in summer (providing other conditions are favorable) and dwindles towards each end of the day and towards spring and fall. This fact is sometimes forgotten by those who are enthusiastic for the substitution of tanks of chemicals for the farmer's broad acres.

It is indeed possible to grow plants under glass and to induce them to manufacture their customary surpluses of food in a very restricted space. But light is a limiting factor; though all the other conditions, including abundance of the raw materials, be satisfied, yet in poor light will the yield be low. The energy of artificial light may be substituted for that of the sun (or added to it); but this increases the cost of food production to a point where the advantages of the method begin to vanish.

Where in this picture is the place of the leafgreen, the chlorophyll? It is certain that the whole elaborate synthesis, even if the proper materials and the energy are abundantly present, fails utterly in the absence of the green pigments. One sometimes finds, in nature or in gardens, leaves that are only partly green, that are mottled or striped or otherwise variegated with white regions. Microscopic examination reveals that the white portions of such leaves are quite normal in every respect except colour; they are albinos. But a simple test with iodine shows that only in the green parts does synthesis of starch from carbon dioxide and water occur.

The manner in which the green pigments accomplish their work is not very clear. The spectroscope shows that chlorophyll absorbs certain kinds of light. The white light that we see in ordinary daylight is a mixture of all the different colours of light in all their infinite gradations.

Chlorophyll appears green to us because it removes blue and red rays from the spectrum and sends to our eyes the remaining parts; the net result being green. The light which vanishes as it passes through chlorophyll furnishes the energy that runs the food-factory in the leaf. A particle or droplet of chlorophyll in the living leaf may be looked upon as a sort of light-trap, in which energy is caught and set to work, as a man-made factory is an arrangement by which the energy of falling water or of expanding steam is cunningly imprisoned and made to accomplish work.

Even when we thoroughly understand the part played by chlorophyll in photosynthesis, we have not a complete picture of the process. It undoubtedly uses one or more enzymes. These are substances used only in very minute quantities and not consumed in processes in which they are involved, in the sense that they do not enter the finished product. No part of the molecule of glucose or starch is derived from any atom of the enzyme, yet the formation of these molecules can occur only in the presence of the appropriate enzyme or enzymes.

The biochemist has not yet isolated the particular enzymes concerned in photosynthesis and is ignorant of their exact nature. When this gap in our knowledge has been closed, we may expect that it will become possible to imitate, in a chemical laboratory, the process by which food is made in a living leaf; as digestion and other vital processes have been so successfully duplicated.

Such is the phenomenon upon which the entire living world depends for its sustenance: a chemical process which occurs constantly in the myriads of green leaves exposed by living plants to the sunshine. Here, in these delicately suspended thin blades of living substance, is made all the food which the plant itself uses and enough besides for birds which eat its seeds, for deer which nibble its young shoots, for man who grubs its underground tubers and for its own offspring to begin life with. Is not nature known for her continual corruption, decay, degradation; in the living tissues of plants and animals, in our own bodies, in all bodies which once had life and are now dead?

These are processes in which the foods and the other organic materials of the living world are returned to the dust or to the air. Only in green leaves does the reverse occur, the manufacture of organic materials from dust and from air; the constant regeneration, recreation, which permits life to spring again from death, which peoples the earth again.

Here in all the verdure of the earth is the meeting-place

of atoms which by taking energy from the sun conspire to give energy to living protoplasm. The resources of the laboratory are far from exhausted and the application of science to industry will doubtless yield new syntheses of food. As I have said, there is no mystery in the making of food—though there are as yet unknown factors, we have no reason to suppose they are unknowable. But whatever the marvelous achievements of the chemist, probably never will it be possible to feed the millions on this earth from the

laboratories. Chemical synthesis will probably remain a scientific stunt and men will continue to turn for nourishment to the green leaves, to the roots and tubers and to the animals which have already converted vegetable foods into their own substance.

Fig. Leaf of Creep Plants

Look about you; even if you live in the midst of congested industry you may catch a gleam of light reflected from the green blade of a living leaf. The iron balconies of crowded tenements have their green gardens. Further field you may look upon wide acres of maize or wheat or the perpetual greenery of forests. Everywhere over the earth are innumerable green surfaces, narrow and broad, thin and thick, downy and waxed, rigid like spears from the ground and trembling pendulous from the ends of slender stalks.

Their numbers can be indicated only by vague comparisons with the sands of the shore and the stars of the night. From each rises a contribution of life-giving gas to the atmosphere of the earth. In each the ultra-microscopic atoms disjoin and swing into new alignments, new unions. In each appear sugars and starches, which pass, a tenuous stream of surplus food, into the trunks of trees or into the roots or tubers or grains or fruits which encumber the earth with riches. Somewhere in all this activity of production are the particular green leaves upon which you are dependent for your sustenance; somewhere the sun is shining intermittently upon your particular green acre, your share in the common wealth and if it were not so you would cease to exist.

Because the source of life was in green leaves, which must grow from soil, wandering tribes settled and flourished exceedingly in the valley of the Nile, so preeminently fitted for the growing of crops and developed a great civilization and arts and sciences, which the Greeks of classic times borrowed and carried forward. Their culture in turn was carried eastward even to the banks of Indus and westward to Tagus and Thames.

Because green leaves made food and because their other-coloured flesh did not, Germanic barbarians invaded Gaul and because of them our western civilization has the shape it has. Again because of the need for land as the source of wealth, farmers pushed over the American mountains, with all their goods and cattle and families, into a savage wilderness and there hacked away the trees and the organic litter of thousands of years and planted their crops and other generations of men plodded into the vastnesses of the prairies, until the great inland empire was won and helped to shape a new civilization.

Chapter 2

Anatomy of Leaves

Most of the green worn by the world is contained in leaves. The pigment is not spread like paint upon their surfaces (which are in fact generally colourless) but is distributed in the intimate recesses of their anatomy. Delicate and thin as the blade of a leaf may seem to our coarse senses,-perhaps three-tenths of a millimeter thick, the eightieth part of an inch—it has a complex internal structure. A leaf is a mechanism.

The manufacture of cloth or paper or steel or glass or any of the multifarious products of human industry depends for its success upon a properly constructed factory; just so do the chemical transformations by which food is made from inorganic materials depend upon properly arranged apparatus—living apparatus.

From the gases of the air, from the watery solutions of the soil brought up through the stem, the proper ingredients are selected and introduced into appropriate vessels; in these they are subjected to the rays of the sun in the presence of minute bodies which contain chlorophyll and the other requisite chemical substances. To build such an apparatus and to introduce into it the right materials in the right proportions and to activate them by the right form of energy have so far baffled even the ingenuity of the modern chemist.

The ultimate parts of living things are far too small to be seen with the unaided eye. To understand something of the mechanics of photosynthesis we must magnify. The microscope is a machine which causes light to travel from the object to be viewed to the viewing eye through a series of lenses, pieces of glass with curved surfaces. In its course the light is bent in various degrees, the rays are spread apart and the several points from which they emanate may be distinguished.

Because of the dispersion of the light, the object becomes dimmer in proportion to the degree of magnification; proper illumination of the object is important. If we place an entire leaf or a considerable piece of a leaf on the stage of the microscope, we see little but darkness. If we play an intense beam of light upon it, we can at best see only the surface; even that, because of its unevenness, is indistinct. To gain a clear picture of the surface layer (the epidermis) of a leaf, we peel it off, lay it carefully in a drop of water on a small piece of glass and cover it with a still smaller piece of glass; then we focus a beam of light from the mirror beneath the stage up through the little piece of leaf-skin into the lenses of the microscope and, finally, into our eyes.

A similar method of illumination is used for the interior parts of the leaf, which we cannot get at in any other way. We cut the leaf into very thin slices (sections), which are laid on their side for study. A series of sections may be run off as slices are cut from a banana and laid in serial order on the glass slide. By studying such a series, the details of which are easily visible at high magnifications, we can penetrate the internal anatomy of the leaf as a whole.

One section is very thin; 10 of them would be about as thick as an ordinary piece of paper and 1000 of them superimposed would form a block not half an inch thick. For purposes of study, therefore, the depth of the section may be disregarded (except at high magnifications) and the section seen in the microscope becomes a picture of two dimensions in the structure of the leaf. Things which seem broken or disconnected may be really joined in the third dimension. Such limitations must be borne in mind in interpreting the accompanying picture, an apparently childish attempt at geometry without any obvious relation to a living leaf.

In such a section much of the leaf is seen to be made of a remarkable assortment of globular, sausage-shaped, or more irregular units which seem to be carelessly thrown together without system or plan.

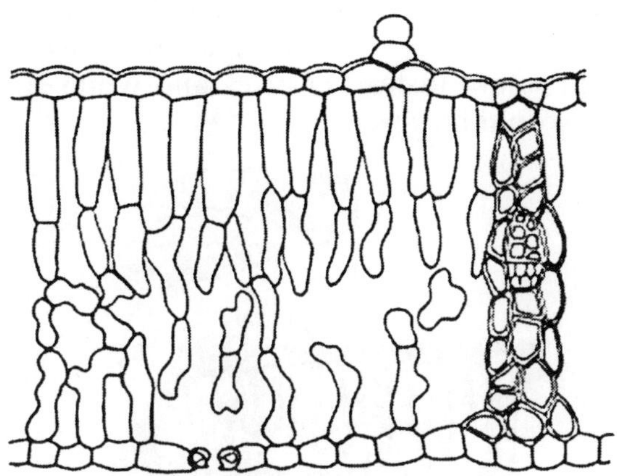

Fig. Cross-section of a Leaf

The upper and lower surfaces may be easily identified; they are formed by layers of smaller, rather box-shaped units. If the section has been cut from a living leaf, many of the interior parts contain minute bodies of a brilliant green colour. These green specks are arranged near the surface of the structural units which contain them; the centers of these units are surprisingly transparent, mostly colourless, apparently vacant. It is perhaps rather disappointing that most of a leaf is water. Water, mounted in a microscope, is transparent; what seemed to be a solid leaf is resolved into an insubstantial tracery of delicate lines separated by comparatively large areas of water.

Our concept is slightly revised if we reconstruct the anatomy of the leaf in three dimensions. The drawing shows the result of putting together a series of sections in proper order. The rounded units of the interior of the leaf are now seen to touch each other at least at certain points and many of them lie in close contact. Although each is little more than a sac of water, the distended solid walls of these sacs together form a continuous firm framework.

Here and there are bundles of tubes, neatly arranged and elaborately ornamented, obviously the veins apparent to the naked eye and forming a sort of skeleton for the more delicate green tissues. The upper and lower surfaces are continuous, thin but firm sheets of transparent material like the cellophane wrapping of a package of sweets or cigarettes. If we could have the picture in colour instead of black and white, most of the interior sacs would be green.

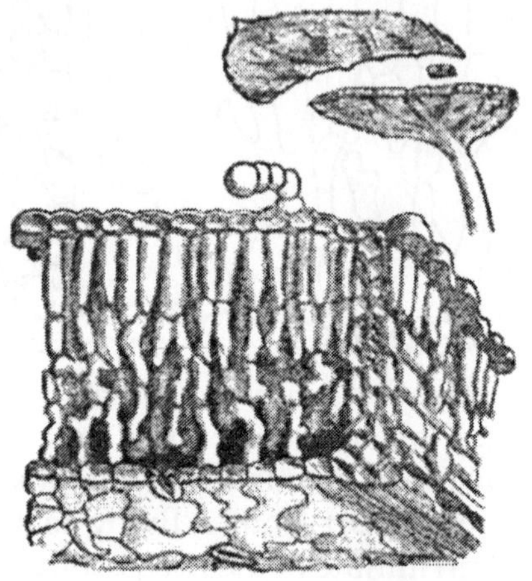

Fig. Leaf of Sun-Flower

The green bodies within them which seem insignificant in two dimensions, are in the aggregate very numerous, sufficient to throw a green mantle over the watery interior of the sac. Everywhere through the leaf, between the water-filled sacs, filling the spaces left by their irregularities, is air; ordinary air in free communication through certain small holes with the air outside the leaf.

Having entered upon a description of the structure of the inner parts of the leaf, we find ourselves involved in microscopic units; which indeed are not peculiar to leaves but occur also in roots and stems and flowers and fruits. Something much like them is found also in the flesh and blood and bones of animals; we ourselves are composed of parts which have much in common with the watery sacs which compose the substance of a leaf. A description of a leaf, if it is at all thorough, is therefore to a certain extent a discussion of the general composition of living things; to do it justice necessitates an excursion. The units which compose all living things, plants and animals, mushrooms, roses, maples, snails, starfishes, moths, mammoths and man,—these units are called CELLS. They were discovered (or at least named) as recently as the 17th century.

The microscope was a relatively new toy then, crude indeed compared with our beautiful modern instruments, but much in vogue as an amusing hobby, a diversion for the inquiring mind. In 1665 an Englishman named Robert Hooke brought out a book of marvels that he had observed with his microscope.

Among other things he had noticed that cork is composed of small pores or "cells" enclosed by walls ("interstitia") of solid material. A similar structure was noticed in other apparently solid parts of plants, such as wood and even in the more tender pithy or green parts; it was evidently general enough to be worthy of comment and to merit description. Hooke, moreover, in his dilatory manner, wrote: "I have with my microscope plainly enough discovered the cells filled with juices...." A greater man than Hooke, the anatomist Nehemiah Grew, described in accurate detail two kinds of components of plants: the fibrous and the cellular, the latter like the "froth of beer."

Through the eighteenth century attention was focused on the walls of cells; only slowly did men come to an understanding that the "juices" within were the actual seat of life, the "physical basis of life" as Huxley later said. In the first quarter of the nineteenth century, not much more than a hundred years ago, the masterly researches of Hugo von Mohl laid the foundations of modern biology; showing that all plants, all parts of plants, were alike cellular and that cells arose from previous cells by their self-division. Similar observations were made upon animals, though in them the cells were harder to see and more mixed with non-cellular material.

It is said that two friends, the botanist Matthias Schleiden and the zoologist Theodor Schwann, comparing their ideas and observations, came to the revolutionary conclusion that the cells of plants are like those of animals, that all living things are composed of essentially similar units. Subsequent research has verified their conclusion. All life is one and man is related not only to the monkey but to the mushroom; all living creatures are built of the same sort of blocks.

It did not occur to the early students of cells that the sapcarrying tubes in the veins of a leaf might also be cells and the fibers of wood and of muscle and of many other tissues. It is known now that not only are all plants and animals made of cells, but that this is true of all their living parts. To be sure, there is much solid matter in a block of wood or in a bony skeleton that is not cellular; it is outside and between the cells, which are imbedded in it. But this solid matter was made by the living inmates of the cells, as the wax of a honeycomb was made by the bees working each in her cell. All living things may be said to be composed entirely of cells and the products of cells.

Hooke's name, cell, has stuck to the unit of life largely as an accident of history. It is not a very good name. It suggests an enclosed space, a cavity and indeed it was in this sense that it was first applied. But since we have found that the functions of life are performed by the minute living bodies within these cavities, our interest is transferred to them rather than to the enclosing walls and the name has followed with the interest. The word cell as used by the biologist now means a small globule of Protoplasm, the living matter within the walls, rather than the space in which it is enclosed or the walls which protect and hold it. Indeed many "cells" are known that are free from walls, that wander unconfined; as if the mendicant friars who wandered over medieval Europe had been known as "cells" because their cloistered brethren dwelt within walls.

Returning to the cells of a green leaf (which are fairly typical of plant cells in general), we see first the bounding walls: thin, transparent, more or less flexible and elastic membranes composed of cellulose. Cellulose is familiar to us in a variety of forms.. Ordinary absorbent cotton is composed of fibers of nearly pure cellulose and cotton goods are woven of cellulose threads. Linen, hemp, wood are cellulose impregnated with various other substances. That familiar industrial product called "cellophane," now indispensable to the merchant, is made from cellulose dissolved and transformed into thin continuous sheets. In fact we shall not be far wrong if we conceive each unit of living matter in a plant as being wrapped in a transparent envelop of cellophane.

The envelop is seamless and usually without folds or wrinkles. Careful research, however, discloses certain extremely minute pores through which the living contents of one cell may communicate with those of neighboring cells. Where cells are in contact, their walls are joined by a common layer of pectic material—the sort of stuff that makes fruit juices jell; this, as it were, cements them together, so that the rounded cells of the interior of a leaf, which appear like so many minute pebbles piled together, really form a coherent mass. This is still more true of the cells which compose the skin of the leaf, which are so shaped that they form a continuous firm membrane above and below and of those which compose the veins, which are rigid enough to form a framework for the more tender tissues of the blade.

The protoplasm within the cell wall—the "juices" of the older writers—is a somewhat glutinous liquid, slightly more opaque than water and seemingly containing many exceedingly minute grains or gobbets of different consistency. It lies in a thin layer against the wall, a transparent membrane, often in motion. Somewhere in it is a rather firmer body, usually flattened on the side against the wall; this little lump of protoplasm, named the nucleus, occupying only a minute fraction of a sufficiently minute cell, has itself an intricate inner structure of the highest importance, which governs the entire activity of the cell.

Since the whole plant is composed of cells and each cell (with a few exceptions) contains a nucleus, it follows that all the nuclei of the plant govern all its life-activities. The same is true of the animal world. We are what we are, we behave as we do, largely because of the nuclei in our multitudes of cells. To the peculiar structure and behaviour of nuclei we shall return in another place.

The layer of protoplasm is also diversified by those firm,

almost solid masses which carry the chlorophyll. If a living cell is observed through the microscope, these bodies are a bright green and only in them is there any green colour. The pigment that is apparently spread so uniformly in nature in the thin blades of leaves is really held in the countless billions of these small biscuit-shaped bodies. They have been considered to be of a somewhat spongy texture, the meshes made of materials similar to those of the more fluid parts of the protoplasm and the chlorophyll in tiny droplets between these meshes.

They are called chloroplasts: "green-formers." They contain yellow and orange pigments as well as green and sometimes red, brown, or blue substances in addition. They float apparently at random in the moving film of protoplasm next to the cell wall. This is one of the most revealing of sights: in the cell of an apparently inert plant leaf, the unending procession of the chloroplasts, small green bodies constantly jostling each other around the cell, up one side, across the end, down the other side, across the end, up the side... without pause and without destination.

Beneath a square inch of surface of an elm leaf there are about 250,000,000 chloroplasts; let us say, at a guess, about two billions to a leaf. Since it is at the surfaces of these bodies that water and carbon dioxide together encounter chlorophyll and are transformed into sugar, the *actual working surface* of an elm leaf only eight inches square is the combined surfaces of all its chloroplasts: about 2000 square inches, 15 square feet.

If we estimate the total number of leaves on the tree, we can arrive at a total of several acres of actual leaf surface; this corresponds to a far greater area of working chloroplast surface—perhaps 500 acres. A single tree is equivalent to a continuous film of chlorophyll of the size of a large farm exposed to the sun and air and supplied with water and the other essentials.

The central (and greater) part of the cell is filled mainly with that very common and essential substance: water. By far the largest part of an ordinary mature cell is water: four- fifths or more by weight. The water is not pure. It contains materials in solution, substances which will be used by the cell in its life activity, substances which have been manufactured by the cell, perhaps inert substances of no particular significance.

The material of this central region, ineptly named the "vacuole" as if it were "empty," is so bound up with the vital activity of the protoplasm which surrounds it that it may well be considered a part of it, sharing in the life of the cell. However, the living processes are so concentrated in the delicate motile layer next the wall that it is convenient and customary to speak of this as the "living" part and to treat the wall without and the water within as "non-living."

Occasional fine threads of protoplasm, like strings of glue or of egg-white, cross the central vacuole, varying from time to time in their position with the movement of the envelop of protoplasm to which they are attached. The protoplasm may flow in such a strand and an occasional chloroplast may be seen taking a short cut across the cell; the almost inconceivably narrow protoplasmic fibre is considerably distended by the passage of the green body, like a python that has ingested a pig. There seems to be always a sharp boundary, a surface film, between the somewhat slimy active protoplasm and the relatively inert watery vacuole.

When we study a section of a leaf we see only sections-slices—of the cells; we see, as it were, a side elevation. Even if we examine living entire cells, as we easily may, the limitations of the microscope compel us to focus on only one level at a time; we see, by changing the focus, a series of "optical sections." The trained observer can put together a series of such sections and reconstruct a cell; the result of such an operation is shown in the drawing herewith.

Here a cell is seen as a transparent closed oblong sac, flattened where it has been in contact with other cells. Contained within the transparent wall is an almost equally transparent layer of protoplasm, slightly granular in consistency, bearing its freight of coloured chloroplasts (white in the picture).

Since these may line the wall on every side of the cell, they may appear to extend completely through its interior; we know, however, that this is not so and that the central region contains only a quite transparent reservoir of water. Red or blue pigments may be dissolved in this region and impart a delicate tint to the entire cell:—pigments quite different in nature and in importance from those in the chloroplasts, which do not dissolve in water. (Sometimes these colours are so concentrated that they mask the green colour of the chloroplasts and cause the entire leaf to have a red, yellow, or brown tint and a special decorative usefulness.) Somewhere in the cell, flattened against one side or occasionally hung in the middle by the protoplasmic strings, is the nucleus, all-important governor and centre of the many activities of the microcosm.

And is this all? Is living protoplasm, the wonderful activities of which have been sung at such length in many a volume, only a sort of thin glue, diversified by a few more solid lumps that float around in its currents?

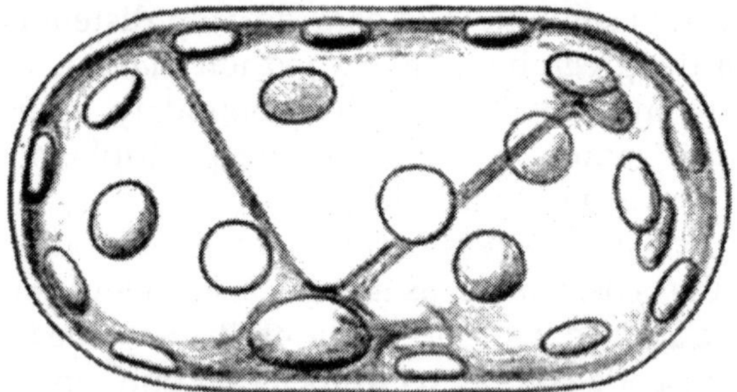

Fig. Microcosm: Unit of Life

One might expect it to be composed of a large number of precisely and intricately arranged parts; one might hope that microscopic study would reveal the difference between the protoplasm of a rose and that of the beetle that devours it. As a matter of fact, if we examine the minute germs from which plants and animals develop, the protoplasm in all is even more disappointingly uniform and unimpressive in appearance.

Chemical analysis, however, has another story to tell. There is no doubt that in protoplasm we have the most complex assemblage of substances known to the chemist. The chemical elements (such as oxygen, carbon, nitrogen, sulfur) of which it is composed can be easily determined and prove to be no different from those which make up non-living objects, earth and air and sea. But these elements are woven together into so intricate a pattern that the chemist, with all the remarkable technical resources of the modern laboratory, is not yet able to duplicate the arrangement and so synthesize protoplasm.

The type of construction, moreover, is one that admits of a multiplicity of variation. As one may make from a small selection of bricks and beams an almost endless variety of buildings, so rabbit protoplasm and carrot protoplasm—having much the same appearance in a microscope and convertible one into the other in the ordinary course of nature—are very different chemical structures. So also we may call to mind that water and gasoline are deceptively similar in appearance; the difference in their activities when they are brought into contact with a flame is sufficient proof of their distinct chemical individualities.

Protoplasm, so surprisingly uniform in its microscopic appearance, lives in a great variety of different kinds of packages. Cells vary in size, in shape, in the thickness of their walls, in their consequent toughness and rigidity and in many other ways. If we examine once more our leaf-section, even in this small portion of one part of one plant we see several different types of cells. Upon the existence of differentiated cells and upon their arrangement in a certain pattern the life and work of a plant depend. We can conceive of a leaf as made up simply of a collection of exactly similar globular cells firmly cemented together; but such a piece of apparatus would not suffice under ordinary conditions to accomplish the chemical work of photosynthesis (to say nothing of the many other activities of living leaves) at anything like the rate with which it occurs in an ordinary leaf.

The principle of specialization is even more obvious if we think for a moment of our own elaborate organization, of the special types of cells in special positions which enable us to see, to hear, to ingest and digest food, to flex our limbs, to perform all our many and pleasant actions. The study of anatomy has its real interest as it illuminates the nature of living activity.

Most of the interior of a leaf, referred to as mesophyll, "midleaf," is filled with rather large cells which contain abundant water and which are characteristically intensely green. Next the upper surface these cells are fairly regular in shape, elongated and arranged side by side like the stakes in a palisade; whence this layer is usually called the palisade layer of the mesophyll. In some kinds of leaves, particularly those subjected to intense light, there may be two, three, or even more layers of palisade cells.

In some other leaves, such as those of shade-loving ferns, there are none. In certain leaves which stand edge uppermost, like those of wild lettuce, there may be palisade layers at both surfaces—the two sides being equally exposed to the sun. It is evident that the disposition and shape of the cells is more or less related to the intensity of the light to which they are subjected; but the details of this relationship are not altogether clear and it would be a mistake to simplify it by supposing that light is the only factor involved in the development of palisade layers. We encounter here a principle of the first magnitude in biology, usually referred to by the very misleading term adaptation: the principle that the structures of living things are such as enable the life processes to continue in a particular environment.

The cells of the lower part of the mesophyll are more irregularly shaped and consequently more loosely fitted; it is particularly through the abundant interstices between these cells that air circulates within the leaf. They are said to compose a spongy layer; a sponge being formed of solid material pierced by numerous continuous passages in which air (or water or other fluids) may circulate.

The veins of a leaf are composed of highly specialized types of cells, which are found also in most other parts of a plant. They are shaped like tubes, mostly very narrow, greatly exceeding other cells in length. There are two kinds. Those which carry water, receiving it from the stem (to which it has ascended through the roots) and delivering it to the mesophyll, occupy the upper portion of each vein, the part towards the upper surface of the leaf.

They have fairly thick walls and taper like needles to each end, where they overlap their fellows. The insides of their walls are often beautifully sculptured and adorned with annular or spiral thickenings. These walls are impregnated with the materials which transform pure cellulose into wood; wood is in fact the water-carrying part of a plant. The living contents of the larger cells may disappear, leaving a continuous passageway for the flowing water.

The lower portion of the vein contains still narrower tubes, thin-walled, blunt-ended, filled with abundant protoplasm. At their ends or sides the protoplasm of one communicates with that of the next by groups of small holes which constitute a sort of minute sieve; the tubes are known as sieve-tubes. Through them passes the surplus food on its way from the factory to the consumers in other parts of the plant and to the storehouses in fruit, stem, or root. The vein is a two-way system of communication, bringing in raw materials and removing the finished product.

The conducting cells just described are often enclosed and bound together by larger cells, which may have thick walls and the whole bundle forms a firm and tough fibre which helps to hold the mesophyll in a flattened form suitable for the reception of air and light. The many veins of a leaf are of many sizes. In the centre runs usually the large midrib, from which branch almost as large main veins; their arrangement varying greatly with the kind of plant.

From these extend the smaller veins, branching and rebranching until all the expanse of the blade is filled with a lace-like pattern, delicate but firm. This skeleton will even withstand the processes of decay and may be found on the forest floor long after the living flesh which covered it has been corrupted and consumed away. Our grandparents, often appreciative of nature in a gently romantic fashion which we are the poorer for having lost, made a fad and hobby of skeletonizing leaves. Books were written on the technique of obtaining specimens and patient artists vied with each other in displaying leaves of beautiful outlines thus reduced to films of delicate lace.

A leaf is internally a tenuous rigid framework clothed with a watery green flesh. We have further to consider the surface layer, the epidermis. In some ways this is the most interesting, the most important detail of the leaf, as our own skin is in our economy. Here is the frontier of living organism, where it encounters the non-living wilderness and through which must pass the materials of commerce upon which living depends.

In the section of the leaf the epidermis appeared to be composed of small box-like cells placed regularly side by side in a single layer. In a carefully prepared and stained section, a layer of wax may be seen to cover the outer surfaces of all the epidermal cells, a layer which is usually thicker on the upper side of the leaf than on the lower and thicker on some leaves than on others; some kinds of leaves have wax in grains or rods instead of in smooth layers. An extremely thick deposit of such material forms the whitish or silvery "bloom" sometimes evident.

Wax is more or less impervious to water; the epidermis is to some extent a waterproof layer. This is a fact of the first importance in the success of the plant. If all the moist cells of the mesophyll were exposed directly to the warm air of a summer day, they would dry out far more quickly than they could be supplied with water by the roots and their life and work would shortly be ended.

Disaster would follow also if the cells of the epidermis could evaporate water freely into the air. Ordinary cellulose walls are quite easily soaked with water and when the cells are alive their walls are usually saturated. At their outer surfaces the water will naturally evaporate and the partly dried wall will take more water from within the cell, much as a wick will draw liquid from its reservoir, by capillary action. But the epidermal cells are sealed and do not lose much water to the outer air.

Evaporation takes place chiefly into the internal atmosphere of the leaf, the air spaces which separate the cells of the mesophyll; these spaces become more or less saturated with gaseous water and the rate of evaporation is consequently low. So clothing hung out to dry does so rapidly if the air is dry, less rapidly if the weather is "humid." The water lost from the cells by evaporation into the air spaces of the mesophyll is replaced from the water-carrying ducts of the veins, which bring it in from the stem, whither it has been conveyed from the roots.

If this were the complete picture, it is evident that after an initial period of evaporation, loss of water would completely cease, the internal air being completely saturated. However, if we examine the epidermis more closely, we find here and there, particularly in that of the lower surface of an ordinary leaf, small gaps, places where neighboring cells are not quite in contact. These are not accidental, not due to tearing of the epidermis or other mishap; they are inherent features of the leaf. Through them water from the humid regions within the leaf passes out into the world and is swept away on the breeze.

A clearer picture of the epidermis and of the openings through it may be obtained by stripping a portion of it from the leaf and mounting it flat for study with the microscope. It is transparent, usually but one cell thick and firm enough so that it may be peeled from the leaf in fairly large continuous sheets. In this view the cells of the epidermis are not box-like as they are in profile; they have wavy edges like the parts of a puzzle of blocks to be fitted together by a child.

The holes or pores just mentioned, through which water escapes as a gas, are much easier to distinguish than in surface view, because of the peculiar shape of the two cells by which each is bounded. The so-called guard cells are crescent-shaped, their ends touching, the concavities forming the opening into the interior. If we return for a moment to the cross section, we may notice the guard cells as they appear sliced across.

They are less in diameter than the neighboring cells and rather more rounded. Their walls are thicker; furthermore they are unevenly thickened-which has a curious result. When the plant is well supplied with water, the guard cells also are well irrigated; in such times they are distended and rigid, the sides erect and curved so as to permit the free passage of air between them.

If water becomes scanty the guard cells (like other cells) may partially collapse; as the internal pressure is diminished, the thick walls straighten and draw together, pushing the thinner part of the wall out as a fold; two such folds meet in the opening and at least partly close it—preventing the loss of still more water by the leaf. The air-pore is technically known as a stoma; it is an automatic and self-regulating valve, which in times of scarcity of water will tend to close and so control the rate of evaporation. A similar closing occurs usually at night also, induced by certain chemical changes incident to the cessation of photosynthesis (the guard cells, unlike the other cells of the epidermis, possess chlorophyll and manufacture food).

The danger of drought is ever present to almost every living plant that grows on land. Its living tissues are composed of water, perhaps to the extent of eighty or ninety per cent and the removal of water not only deflates them but induces chemical and physical changes which may result in death. Water constantly evaporates into air and must be supplied from the substratum, where the supply is variable and dependent upon factors outside of the plant's control. The quantities of water which pass through a healthy plant are surprising, scarcely credible. A grass will evaporate its own weight of water in a day. A maize plant will lose in the same time about a gallon of water and an acre of maize will lose during the growing season the astonishing total of 300,000 gallons.

A large tree drinks up an enormous amount of water from the soil and pours most of it out through its millions of stomata: perhaps a couple of thousand gallons during the summer. It is small wonder that the moneygreed which destroyed the forests over so much of the United States (and is still destroying them) has changed even the climate of vast stretches of country.

The rain that falls, instead of being lifted up again into the air to humidify it and fall yet again, pours its destructive flood down the riverchannels to the oceans, carrying precious topsoil in its current. In regions which have been deserts since olden days, the palms and other trees are mostly confined to the borders of springs and pools. A single date palm may evaporate into the hot dry air 190 gallons of water *in one day* and runs up a deficit of 35,000 gallons in a year; sufficient reason why this ancient staff of life does not invade the arid sands which hem in the narrow oases.

When growers move plants or set cut branches in wet sand to make new roots of their own, they frequently remove part of the foliage. The value even of such a sacrifice of food-making area is evident from the facts just detailed.

Since the life and well-being of a living plant are so continuously endangered by the tendency of water to evaporate into the air and since we expect plants to be structurally suited for successful life, it has occurred to many students to wonder how it is that land plants have not invented sealed leaves from which no water can evaporate.

Nature has equipped plants and animals with so many self-regulating and protective devices; how does it happen that this vital problem of plants has been overlooked? Marvelous automatic mechanisms as are the stomata, would it not be still better if they did not exist at all, if the epidermis were completely closed and waterproof? In a world of creatures which fit their environment, which are adapted, is it not monstrous that such a badly managed business as the death of plants through drought should be allowed to occur?

The difficulty is that the carbon dioxide of the air, so important in the food supply of the plant and of the world, must *enter* the living tissues if it is to be made into sugar. The epidermal cells are continuously bathed in air, of which carbon dioxide is a part. But no gases can enter a waterfilled sac except by first ceasing to be gases; by dissolving in the water. It is a cardinal principle of vital physiology that (with minor exceptions) *all substances that enter or leave living cells are dissolved in water*. Since the outer surfaces of the epidermal cells are well waxed and to a large extent waterproof, they do not offer to the carbon dioxide a watersoaked spongy surface in which it can dissolve. The absorption of this raw material can take place only from the internal atmosphere of the leaf at the moist surfaces of the cells of the mesophyll—the same surfaces from which the water so readily evaporates. This internal air must be continually replenished with carbon dioxide from the outer air, if photosynthesis is to continue. As carbon dioxide enters the leaf, water is lost; each gas moving in accord with the laws of diffusion towards a region where it is less concentrated.

Anything that interferes with the loss of water likewise must cut down the rate at which food is made; a plant fully waterproofed would starve.

One result of the coat of dust which disfigures the plants of our roadsides is to clog up their minute pores, to suffocate them (as well as to cut off much of the light from their mesophyll). The soot deposited by cities has a like effect, though it does even more serious damage by releasing injurious acids.

These curious small slits in the leaf-skin are therefore more than details of anatomy; they are the portals of life, the gateways through which flows the supply of carbon for the manufacture of most of the food of the world. Their minute size may seem inadequate for their gigantic job. Each is smaller than the hole made by the finest needle: in the neighborhood of 0.01 millimeter (1/2500 inch) long and, when fully open, about half as wide.

But what they lack in size they make up in number. There may be 50,000 in a square centimeter of leaf surface (over 300,000 in a square inch) on the under side alone. An ordinary geranium leaf may be pierced by two million stomata; the entire plant may have twenty millions. The number in the two or three acres of leaf surface displayed by a large elm requires more digits than the imagination can assimilate.

The stomata are often spoken of (in textbooks and other compilations of misinformation) as "breathing-pores." The plant performs no function which can properly be compared with the inhalation and exhalation of an animal and, while the stomata indeed serve as a means through which gases are exchanged between the living cells of a plant and the air around it, to speak of the plant as "breathing" obscures any true picture of its activities.

The evaporation of water from leaves is so important that it must be reckoned among the vital processes of the plant, among the activities which together constitute its life. This process has been named transpiration. As soon as we label anything a life activity, as soon as we give it a name, some one begins to seek for the benefits which it must confer; so unthinkable is it to many persons that something can occur normally in a living creature without being in some way essential to its continual welfare.

Various supposed benefits of transpiration have been proposed by those who think the plant does it on purpose. Careful research has failed to disclose any real advantage of the evaporation of water, while the disadvantages are obvious. It is necessary to regard it as simply a natural process which occurs (like the other life activities) because the leaf is constructed in a certain way and without reference to the desires and needs of the plant. This is after all simply the scientific point of view—to regard nature as natural.

Thus far we have dealt with the anatomy of leaves in their common forms. Such leaves as those of a lilac bush or an oak tree or a sunflower plant are so constructed and the majority of familiar leaves which grow upon trees, shrubs and herbs in our fields, woods and gardens. The differences in shape, in size, in the teeth along the margin, in the hairs that often cloak the surfaces, these differences seem unimportant to the life of the leaf and of interest only to the systematist who must identify plants and to the artist who must represent them.

Doubtless our friends who see a felicitous arrangement in every detail of the organization of a living thing must logically conclude that a plant grows hairs on its leaves in order that it may be named by a botanist.

There are also other types of leaves, which live lives of some peculiarity. To live under water is as difficult for an ordinary plant as for an ordinary mammal; but there are plants which grow normally in this medium, as there are animals fitted by nature to such an environment. The leaves of such plants have structures as different from the ordinary as their problems are different. For them transpiration is not a factor, nor is drought a present danger. Their leaves are—as might be expected—innocent of waterproofing layers, of an inner system of air spaces, of stomata.

On the other hand, they live under very poor illumination. If they were as thick as an oak leaf, the upper cells would effectively mask the lower and entirely rob them of light. It is not surprising, therefore, that we find fewer layers of cells in aquatic leaves. Sometimes they are only a cell or two thick and the blade is dissected into hair-like parts whose every cell is directly exposed to what light there is and to the water from which the carbon dioxide (in solution) is taken.

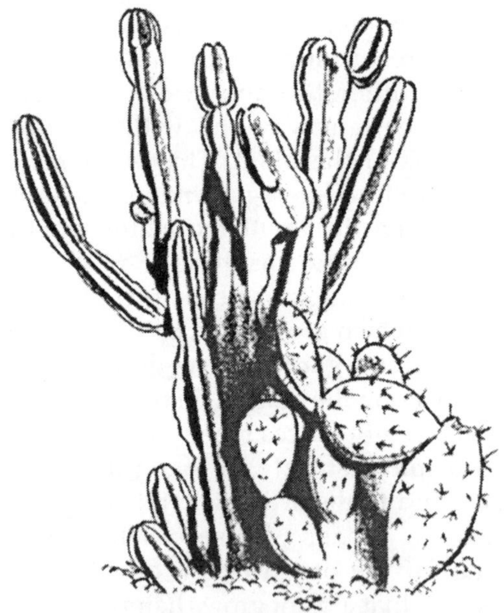

Fig. Cereus Opunita

In the seemingly waterless desert also plants grow. Where it appears that there is not water enough to fill a blade of grass, where men die of thirst,—even in such regions do certain plants live and thrive. They are unlike the plants we know in temperate or in tropical fields and forests. A journey to the sandy or stony wastes of southwestern North America acquaints us with a plant world of wholly different aspect. Cacti of fantastic shapes sprawl over the hot earth, or rear their ungraceful bulk above it; fleshy green stems on the tips of which small leaves may be formed in spring but almost at once fall away.

In these plants the manufacture of food is carried on in the stems, parts which are not so subject to drying as are leaves. A ten-foot cactus may lose only an ounce of water in a day,—in contrast with the several gallons lost by a ten-foot apple tree. The stems of cacti, curiously enough, often resemble leaves. The common prickly pears seem to be chains of fat leaves, each arising from the apex of the one below.

Fig. Agave

Their anatomy teaches us that this is not a correct way of describing them; that parts with the internal structure of a stem may have the external appearance of a leaf--and vice versa. One of the common pastimes of the natural scientist is thus exemplified: to propound that what looks like a leaf is not a leaf, what looks like a seed is not a seed, what looks like a bug is not a bug,—in general that things are not what they seem.

Other desert plants have stiff and sharp leaves which rise like bayonets from the ground. A section through the thickness of such a leaf shows layers of large cells packed with water and with acids which have the power of "binding" the water, preventing its evaporation; these cells often lack chlorophyll. The plant pays a price for subsistence in such unfriendly regions, a price in reduced facilities for making food. Such plants in general are referred to as succulents.

Among them are those small inhabitants of our greenhouses and window-sills which have become household pets. Their leaves are pale with wax and often as thick as they are broad or long; they grow close to the soil on a stunted stem, often arranged compactly in beautiful geometric precision. Some are almost buried in the hot dry sand, only the transparent ends of their truncated thick leaves exposed to air and light.

Some lie mixed with the burning stones which they so resemble in shape and colour that one can scarcely discern the presence of the living thing. After the brief and infrequent rains these thwarted and stunted plants put on new colour. Beautiful flowers appear with what seems magical rapidity, covering with their bright colours the squat leaves which make their growth and glory possible.

It is a far cry from these plants, which seem to be rather mineral than vegetable, to the six-foot leaf-blade of a banana tree, to the great floating platters of the giant waterlily of the Amazon. Yet in all this diversity of structure certain features are constant. The cells are no bigger in the waterlily than in the stone-plant.

The colour is much the same in all leaves (chlorophyll is present even if it is obscured by red, yellow, or brown pigments), all are exposed to the same gases (whether in air or dissolved in water) and to the same sunlight and in all except the white leaves of a few parasites some sugar is made from inorganic substances.

These manifold objects of beauty which greet our eyes and inspire our art, varying in shape, in size, in colour, in indument, in succulence,—all are essentially so many sugar factories. Sugar, beloved of children, consumed by the millions of tons, commercially obtained from only a few crops, has its origin in all the green cells of plants, where it is formed out of nothing more recondite than air and water.

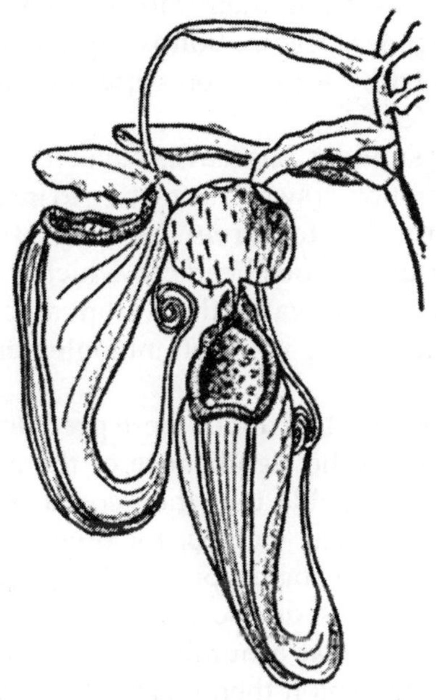

Fig. Nepenthes

Among the weirdest of leaves are those which shape themselves into vases, cups and pitchers. Their cavities become filled with water and in them numerous small insects meet their end. Some form sweet secretions around the treacherous lip of the cavern. The hapless intruder is lured into the tempting interior and slips down into the water, prevented from crawling out by down-pointing bristles, by slippery overlapping scales of wax, or by other such arrangements.

There is evidence that some of these plants pour digestive juices into the liquid of their hollow leaves, which convert the drowned animal bodies to soluble foods; if this does not happen, ordinary bacterial decomposition will quickly reduce the small corpses to the same condition. The dissolved products of digestion are said to be absorbed into the plant tissues as they are from a stomach into the tissues of an animal. There are other carnivorous plants, surprising us with a habit that seems rather to belong to the animal world. Dionaea has for leaves efficient fly-traps: small blades which fold like the two halves of a book along their midrib, the edges garnished by interlocking spines. The motion is rapid enough to entrap a fly and is caused by a touch on certain delicate bristles in the centre of the leaf.

Drosera, the sundew, has leaves covered with erect sticky tentacles. The secretion is sweet, attractive to insects, who come to eat but are unable to leave. As they struggle to free themselves from the viscous heads to which they have stuck, other tentacles arch in towards them from all parts of the leaf and rapidly complete the work of entanglement; after which comes digestion.

Much has been written about these predatory plants; little is known even yet of the significance of the animal addition to their diet. It is probable that they could live without the digested remains of their victims, for they are green plants, their cells manufacture sugar apparently in the same way as less bloodthirsty leaves do. So great is their appeal to the popular imagination that a number of irresponsible writers are only too glad to exploit their charms; only today in a well known newspaper I find a page or two of absolute nonsense on this subject.

It is still usual for uninstructed persons to refer to animals as "living" and to limit the concept to them; plants are not "life." A religious sect of India admits that onions are alive; the exception betrays the rule. It is true that plants, even carnivorous plants, seem to us relatively inert beside the swift motility of an antelope or the terrible energy of a tiger; beside even the scurrying and tail-wagging of the household dog. It is difficult to form a mental concept of the *chemical* activity within a motionless plant leaf, activity which is silent, which involves the motions of swarms of molecules rather than of the entire mass; activity essentially similar to what goes on in our own bodies, masked by the more apparent motions and changes.

We are obstructed in our understanding of life by the small size of the living apparatus. The microscope aids our eyes but our imagination does not always keep pace. Suppose that by nibbling the appropriate side of the right kind of mushroom we could make ourselves as small as we wish; say about one-hundred-thousandth part of our present clumsy size.

If we could then walk out on the stalk of a leaf we might climb precariously among the hairs of the lower surface—which would seem to our littleness as big as the great vines of a tropical forest—and reach a stoma. The opening would seem to us about five feet long and two-andone-half feet wide and we might squeeze through. Inside we could stand erect in a more marvelous and beautiful cavern than any yet discovered by man. The air would be warm and humid and the light a soft green.

Fig. Victoria Regia

On every side the glistening rounded surfaces of great cells would tower a hundred feet over our heads, where we could dimly discern the hanging green stalactites of the palisade layers. In the cells we should see green bodies about the size of dinner plates, many of them in motion. Before our eyes, in awful silence, starch grains would increase in size, grains as big as footballs.

On every hand dim and tortuous passages would wind away between the cells of the sponge, spaces in which we should stumble and slip on the moist rounded surfaces as we clambered around. Here and there we should encounter large pipes lying side by side in bundles, sometimes completely obstructing our way; the largest of these pipes, through whose transparent walls we might just discern the water flowing within, would be perhaps six feet across in our eyes.

In the smaller ones we might be able to see the glutinous protoplasm moving, in whose meshes the food from the neighboring cells was being carried away. If we had the means of cutting into these vessels we should not lack for food. The air would be rich in oxygen, which might compensate for the high relative humidity.

By such a mental journey may we perhaps penetrate to the strangely beautiful and quiet factories where is made the food of the world.

Chapter 3

Architecture of Plants

Outside in the chill sunlight of early March the gaunt trees uphold wooden fingers against the pale sky. A branch fell in yesterday's wind; it will soon be cut up and burned on the hearth. It is difficult to believe that that stick, hard, fibrous, combustible, was recently alive; that in it was living material like that in your own flesh and blood,—the whirling atoms of life.

Yet will those stiff trunks and branches, in a few weeks now, break out a delicate greenery, visibly surging into new growth; adorning themselves also with flowers according to their kind. In a short time the entire brown land clothes itself with the voluminous cloak of green which it wears for the summer.

Life, now sluggish, dormant, suddenly quickens and moves towards abundance and fecundity. Later, after the glorious apotheosis of the fall, the denuded branches will stand there again, a trifle more extended in space, still hard, stiff, apparently nothing but potential fuel:—but still alive.

The trunk, which seems to be but an inert supporting column, is alive, as the leaves and flowers and fruits are alive. If we were for a brief time blessed with sight so penetrating that we could see through the bark and into the wood, so microscopic that we could distinguish its elements, we should revise our indifference to its activity.

We should see water in thousands of minute tubes passing up the plant, moving rapidly enough so that occasional air bubbles carried along in the stream would flow up with visible speed. We should perhaps be able to discern just beneath the hard outer surface a region of very minute tubes filled with glutinous protoplasm in rapid motion; in this part food passes from the leaves downward during the season of active food-making and up again into buds and growing twigs and developing fruits at the proper times. If some power should also be vouchsafed of distinguishing foods in solution from their surroundings, we could watch the sugars and other foods pouring into the living cells of the wood, accumulating there in certain narrow reservoirs; later, perhaps, to be tapped and drawn off by us for our own purposes.

In some trees we should discover canals in which resin moved on its viscous way; in others again we should find tubes containing the latex so desired of a rubber-tired and gum-chewing civilization. If we extended our view over a long period and made a careful series of measurements, we should be able to see that the trunk is a self-expanding mechanism, adding to its girth most mysteriously from within, cracking the hard surface which becomes furrowed and flakes off and which is unceasingly replaced by a new surface.

In some kinds of plants the old trunk gives further proof of vitality by putting out new branches, covered with green leaves or with flowers; this is especially marked after periods of drought or when other factors have disturbed the normal life of the upper parts of the tree. Indeed, if you fell the tree, leaving only the hard and apparently lifeless stump rising from the soil, it may surprise you by bursting into abundant growth and in a short time is covered with a bush of green twigs, asserting its dominant will to live.

The woodworker knows well the different grains of the different kinds of wood: the light soft wood of white pine has its fine straight grain, the oak has large pores and bright lights revealed by certain methods of cutting, maple has scattered small pores and shining flecks. All these signs speak to him a particular language, telling how they will split, how they will take and retain a polish or a stain, how they will warp or crack when the atmosphere changes; he reads his woods as quickly and as surely as you read the printed page—almost without knowing how he does it.

The botanist probably lacks the craftsmanship which can deal with wood and fashion it to his choice; but he is nevertheless able to penetrate its nature in a way that is usually denied to the craftsman. Botanical technique reveals a beauty scarcely suspected by one not initiated into microscopy; besides providing a fuller explanation of the qualities of the wood known to the woodworker.

Wood—xylem is its Greek or botanical name is microscopic in texture. The microscopist cuts it into slices thinner than this paper and places them on the glass slides of his magnifier, illuminating them from below like any other thin section. He sees the pattern of "cells" that led Hooke to his first statement of the structure of plants. The arrangement of the cells and their shapes vary in different kinds of trees; also with the direction in which the cut was made. Wood is made mostly of very long cells, their long dimensions parallel to the length of the trunk or branch.

Cut crosswise, these tubes appear angular or somewhat rounded, approximately equal in all diameters. Cut lengthwise, they are but parallel lines with spaces between, the spaces here and there intersected by transverse or oblique partitions; the tubes are so long that both ends of them do not usually appear in the field of vision of the microscope. The side walls are usually adorned with thickenings placed in a definite pattern, in rings or spirals or in transverse bars; or with round and curiously formed pits in an otherwise uniform surface.

Attentively we scrutinize these wooden tubes; not only do they contribute to the beauty and to the strength of our furniture and dwellings—they are also the channels through which water passes from the roots to the foliage displayed above. The thick-walled upper halves of the leaf-veins are but frayed ends of wood which penetrate the leaves.

The earliest anatomists (more familiar with the breathing-tubes of some animals) thought that air passed through the tubes of the wood and were much exercised over the possible usefulness of the rings and spirals which line them internally. Even now that we know the true function of the vessels we are uncertain of the significance of their structural adornments. There is a great variation in size among these wooden vessels; some of them are eight or ten times the diameter of others in the same region, consequently sixty or one hundred times as large in the area of their cross section.

In many kinds of wood there is a definite gradation of different sizes of cells. The largest are those formed in early summer, when growth is most vigorous. As the season wears on, the cell size becomes progressively smaller (for new cells are added to the wood during the summer, in a way to be explained later), until finally growth ceases for the year.

The next spring brings a sudden resumption of activity and the immediate production of large cells which contrast sharply with the small ones which preceded them. The line along which the cell size changes abruptly is usually visible even with the naked eye and is the basis of the "growth rings" familiar to one who has felled a tree or looked at a stump.

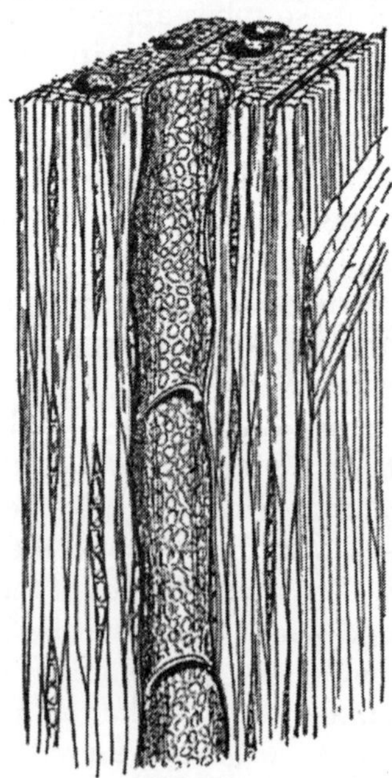

Fig. Wood

The older parts of the wood are those nearest the centre

of the tree. As their age increases, their cells die and their cavities often become closed by the intrusion into them of growths from neighboring cells. They may also become discoloured and hardened, so as to form the "heart-wood" often valuable as timber. This is indeed little more than a dead supporting column in the heart of the plant. New "sapwood" is continually formed around the older wood, just under the bark. In this the sap runs upward, the water and its freight of mineral salts from the soil and various living cells manifest their activity.

Even in the living wood the water-tubes themselves, the conducting vessels, are dead; their protoplasmic content, which built up the thick walls with their geometrical ornaments, has disappeared, its work done. Mixed with these waterfilled cavities, however, are many living cells; the sapwood is much more than an inert zone of fibrous and conductive tissue.

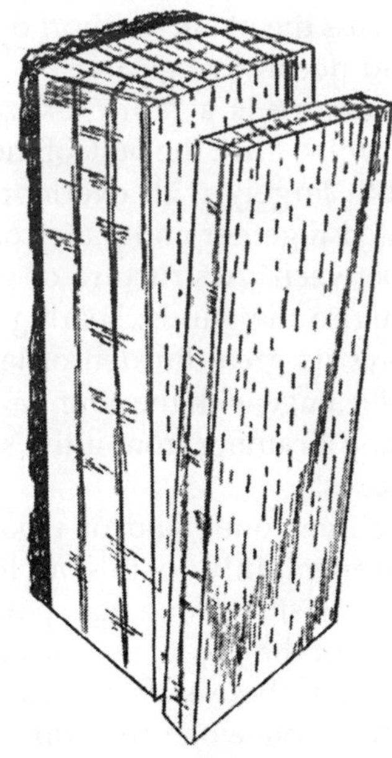

Fig. The Dead Wood Log

The woodsman knows also that the growth rings do not

compose all the figure of the grain; there are rays, which extend from the circumference towards the centre. These are made of cells which also are elongated, but in a direction transverse to the length of the stem. They are largely living cells, not merely shells surrounding columns of water. The distinction between the living rays and the largely non-living vertical components of the sapwood is demonstrated by some forms of rot. The fungi which cause the rot are able to attack the dead wood, the sap-carrying tubes that extend up and down the trunk, but leave untouched the living cells of the rays which cross them. In the crumbling mass the rays are like thin blades extending from the periphery towards the centre, each only a few millimeters high.

If from a sound trunk a board is sawed parallel to a growth band, the rays are cut across; they form the innumerable narrow flecks or short lines in such a cut. If the cut is radial, intersecting the growth bands at right angles, these form long straight lines, while the rays, being split, form bands extending across the plank for short distances; they have often a peculiar and decorative gleam.

The entire structure is a fabric of interlocked fibers: longitudinal fibers which form the bulk of the wood, of various sizes and shapes but largely with overlapped pointed ends, all strongly cemented together and plates of transverse fibers inserted radially between. A structure of great strength, yet flexible and yielding to the winds; bearing aloft and holding out to the sun's rays its great burden of leaves, at the same time supplying that leafage with the abundance of water which keeps it in health and draining from it the superfluity of food which it manufactures.

In temperate climes one band of wood is formed each year, its limits often plainly visible as long as the trunk endures. On a standing stump one may polish the cut surface and trace the history of the great tree back through the years, through hundreds, even thousands of years: this ring was the outer surface of the wood when the war between the states broke out; this was formed while the Declaration of Independence was being, signed; the tree had grown so far when Columbus set foot for the first time in the New World; here is the ring formed the year that William the Norman crossed into Saxon England; here, far in the interior now, are rings, small in circumference, that grew while the Galilean walked on earth.

Some rings are broader than others, some much narrower than the average; these may mark years of abundant rainfall or of drought. Cycles of climatic change may thus be identified and verified by reference to old records. Forest fires or lightning bolts left their marks; the bark healed them over and fresh wood was formed outside them, preserving the record for us to find. Having dated particular bands, recognizable by some peculiarity of arrangement, we may find the corresponding peculiarities in stumps or logs or beams of unknown antiquity and so work forward to the dates at which the trees were felled.

In such a way we have discovered the time when Indians of the southwestern states cut the timbers which they used to support their roofs. The method may be pursued, from a dated timber to the corresponding signs in an undated one, almost indefinitely back through man's structural relics and provides an important tool for learning their antiquity.

How is it all formed? Whence come the new bands of wood, appearing one outside the other every year? The answer lies in the bark. The bark seems at first sight even more dead than the wood; a rough, dark, tough substance constantly being split or flaked off by the expansion of the tissue within. But every boy who has made a willow whistle knows that this is not a true picture. The outside indeed is dry and lifeless, but the inner layers are tender, moist, often green: obviously alive.

The bark is not one thing but many: successive shells of entirely different tissues. It has a single name only because it may be stripped from the wood as if it were a single layer. Much of its bulk is formed of cork and is in truth dead. The cork is much the same, chemically, as that material with which we stopper bottles; it is formed of cell walls impregnated with a fatty material impervious to water, the cavities of the cells containing air instead of protoplasm. The fine grades of cork, uniform in consistency and sufficiently resilient and fine-grained for cutting and shaping, are derived from the bark of certain trees; particularly the cork oak of the Mediterranean shores. At first a stem has a living skin, an epidermis like that of a leaf. Beneath this layer, patches of cells begin to divide, multiplying their number; the patches spread, join, become a continuous cylinder.

The new cells of which this cylinder is composed begin to manufacture fat-like substances, which are deposited in their walls. As they surround themselves with this barrier to water, they cut themselves off from intercourse with the living world about them. No water can enter them from within the stem, no foods can enter; they die. Their compressed and stratified remains form the corky outer layers of the bark. The epidermis and other cells outside them die also, for the same reason: the new cork separates them from the stream of life within the stem; all that is left for them is to evaporate, wither and pass away.

The formation of new cork is continued by a like process. New zones of dividing cells constantly appear in living tissues beneath the older ones and form new zones of cork. Thus is a thick waterproof jacket built up around the living cells of the interior bark, a jacket which is always being added on the inside as it wears away on the outside.

Within the corky layers is a thin membrane of living and very active cells; named the phloem. It is composed of very long cells, very narrow, thin-walled, densely filled with living protoplasm which is usually in motion. These tubes communicate with those above and below and to each side of them by means of minute perforations in their walls; holes which occur in groups so as to form a sort of sieve; the tubes of the phloem are those sieve-tubes already mentioned as forming the lower part of a leaf-vein.

Their importance is briefly stated and easily understood: through them flows the stream of surplus food made in the leaves and carried to all other parts of the plant. All the food in roots, tubers, fruits, seeds, in any of the convenient packages in which we find food stored for our own use, all has come there through sieve-tubes. Consider the food which you have just eaten; an essential constituent of it began its animated journey by entering the stoma of some leaf somewhere in the world; this encountered other elements which had traveled through xylem to meet it; the product of their intercourse accomplished part of its long voyage to your plate through a series of sieve-tubes.

The phloem, although constantly added to by the continuing growth of the stem, is always a thin layer, a mere delicate living skin on the inner surface of the bark. Its older regions (the outer layers) are constantly being involved in new cork formation and passing out of the roster of living phloem cells. Its thinness and its position near the surface of the tree are a source of danger to the plant.

Anything that cuts the phloem cuts the lifeline of the parts below; they will starve. When the roots die they will decay and the upper parts in turn will die from lack of water. I have seen a field dotted with dead trees, each carefully girdled—a ring of bark removed down to the wood—by the owner of the land; a farmer too poor or too lazy to fell his trees, who in an hour or two cut their lives across and was content to wait until they died and were blown down by a spring storm. Rabbits who love to gnaw the succulent bark of saplings likewise may steal their lives.

Those who use trees as posts to which to fasten guys may lose their trees after a few years; the wire cuts down into the bark and strangles the stem. If the plant manages to live, evidence of the trouble within may be seen in the swelling which is developed just above the constriction; caused by an accumulation of food which cannot pass down so fast as it is supplied and which stimulates the gorged cells of that region to overgrowth.

Finally, within the phloem, we come to the layer which answers the question raised: Whence does all the growth come? This is another cylindrical sheet within the cylinder of phloem; cells which have the remarkable power to form new cells by dividing themselves. Imagine a cylinder of material thinner than paper which can automatically cleave itself into two cylinders, one within the other. The inner one enlarges, thickens, becomes wood, stretching and spreading the outer cylinder and all the other layers of cells outside it; the outer cylinder repeats the cleavage, the plane of division again being parallel to its surface.

Or the outer of the two concentric layers may mature; it becomes phloem, the inner one retaining the divisibility which characterizes this region. So it continues; new bands of xylem are added on the outside of the old ones, new bands of phloem on the inside of those already present. The vast trunk, so many years in the making, composed of so many billions of billions of cells, is a deposit of the living skin which surrounds the wood and lines the bark.

The trunk will heal itself when wounded. If lightning makes a gash through the bark and into the wood—examine the riven edges a few weeks later; you will see an advancing rounded mass of smooth hard cork coming from the inmost living tissues of the bark, creeping out over the naked wood, covering it slowly with fresh growth-layers and food-carrying layers. Cut off a branch and in due time the bark creeps inward from all sides, covering the cut end, stopping the loss of water and—with luck—preventing the entrance of the germs of rot and disease. Perhaps new shoots may sprout from the callus and several branches flourish where but one stood before.

Self-healing, constantly increasing its girth from within, carrying water and food in two systems of tubes, accumulating surplus food in hidden places, manufacturing gums and resins, putting out new foliage and flowers year after year:—the trunk of a tree is a very complete and complex piece of machinery; beautiful to all who can see its grace of line and its strength, still more beautiful to those who can sense something of its inward organization and its vigorous activity.

What of plants that have no woody trunks, herbaceous plants with green and delicate stems? Have they conductive systems, mechanisms of continuous growth? They are covered with living epidermis, they are composed mainly of thin-walled water-filled large rounded cells much like those of a leaf—the outer ones have chlorophyll, make some carbohydrates. But they have also wood and bark, xylem and phloem, arranged in a definite pattern.

The conductive cells compose strands or bundles like the veins of a leaf, strands that extend more or less vertically in the stem. In a transparent shoot one may see them merely by holding it up to a light; it is a simple matter, moreover, to let these strands drink up a coloured fluid (which passes up through the channels normally taken by water), when they become plainly visible. Other stems must be cut crosswise or lengthwise that their conductive apparatus may be seen. In cross section the strands are found usually only a short distance beneath the epidermis, making a circle or occasionally a square.

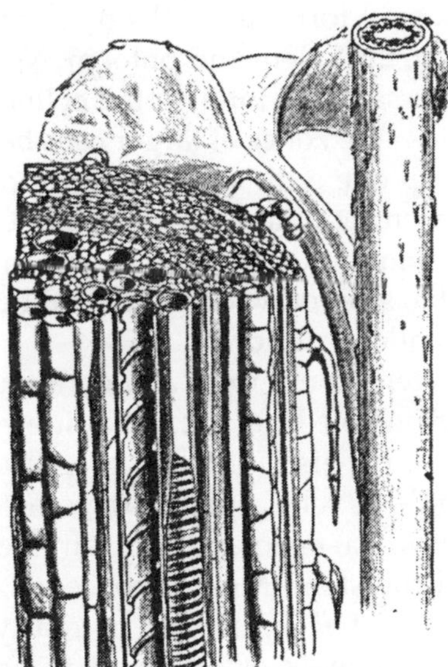

Fig. Helianthus Stem

If we study one strand through a microscope we see at once the cut ends of the water-carrying vessels and the other thick-walled fibers which make up xylem; to these small patches the wood of such a stem is limited.

The wood lies on the side of the conductive strand towards the centre of the stem; towards the epidermis we find a narrow zone of those very small cells which carry food, the phloem; appearing polygonal or rectangular in cross section. Frequently the phloem is bordered on its outer side by a strand of fibrous cells, thick-walled.

The bulk of the stem, including all that part which lies within the circle of conductive strands, is composed of watery thin-walled cells, in which foods or wastes may accumulate; an undifferentiated tissue of no particular significance in the life of the stem (indeed often the central part, the pith, dies and may entirely disappear, leaving a cavity),—yet alive, capable under certain conditions of growth and reproduction.

Even in such stems there may be increase in diameter, brought about in much the same way as in a woody stem. Between xylem and phloem in each bundle of conductive cells is a region which may form new cells by division; these new cells being added to the xylem and phloem. Nor is the cylinder of new tissues necessarily an interrupted cylinder; the impulse to divide runs across the zones of larger cells between adjacent bundles, linking up one with another and forming new cells in continuous concentric cylinders just as in the trunk of a tree. A tall sunflower stem, at the end of its growing season, may have a considerable strip of wood concealed beneath its green and apparently tender exterior.

Such stems often do not form any cork on their surfaces, but retain for their whole brief life the epidermis with which they started growth. This is a layer of cells much like the epidermis of a leaf;—its cells usually to a greater or less degree covered with wax which enables it to retain the water flowing so abundantly in the stem. Most herbaceous stems also are more or less hairy. The hairs are of many types-single protruding cells, filaments of cells end to end, or massive structures composed of many rows of cells.

They may be matted like wool or lie uniformly flattened in one direction along the stem; they may be soft or hard, may be terminated by a swollen gland which exudes fragrant oil, or by a needle-sharp point. They may be long barbed prickles, or they may have bulbous bases filled with poison which is released into an intruder's skin by the breaking of the fragile tip. You may caress the softly downy stem of a geranium and enjoy the aromatic odor which the broken cells emit; a similar treatment of cactus or stinging nettle is not to be recommended.

Stems of common grasses, including the cultivated grasses corn and wheat, barley and oats and rice and the rest, differ from those of beans and geraniums and sunflowers in the arrangement of their vascular strands: these being scattered throughout the stem instead of lying in a cylinder. They are usually more numerous towards the periphery and all the cells around these outer ones become thick-walled and woody as they grow older; everyone knows how tough an old corn stalk is.

The structure of each individual bundle is comparable to that of a sunflower; but there is usually no increase in diameter after all the cells of the stem reach full size; there is no tissue which continually adds new cells after the primary structure is once organized. The basal parts of a corn stalk have about the same diameter as those near the tip, although they are older. A palm is a member of the same alliance of plants; its cylindric shaft contrasts strangely with the tapering trunks of elms and maples.

Plants, whether trees, shrubs, or herbs, resemble the tallest man-made buildings in one respect: much of them is beneath the ground, unguessed at by the casual observer.

An old root has much the same appearance as an old stem, whether seen in the mass or dissected. Indeed it is sometimes difficult to distinguish short lengths of stems and roots. We find the same solid core of wood, surrounded by a thin coat of bark. Each of these parts of a root has about the same structure as those of a stem, even in microscopic detail. The roots and the stem are parts of one continuous system, which must not only support and anchor the leafy top but carry water up to it and remove from it the surplus food. A young root, however, differs quite strikingly from a young stem and the differences bring to mind some very pertinent questions.

The part of a young root that has just matured into recognizable tissues out of its amorphous embryonic condition possesses a conductive system of cells which lie close together in a small central core. This occupies only a small fraction of the cross section; between its boundary and the epidermis is a large expanse of thin-walled cells like those of which the pith of a stem is made, filled with water and often containing large quantities of surplus food.

The epidermis is a much more delicate tissue than that of a stem and its hairs are tender thin-walled threads that penetrate between the small particles of soil with which the root is surrounded. The conductive core, the central cylinder, has usually several strands of wood which alternate with as many strands of phloem; the strands of wood usually become joined and continuous in the centre of the root, so that in cross section they form a single band or a three-pointed figure or a cross or a star.

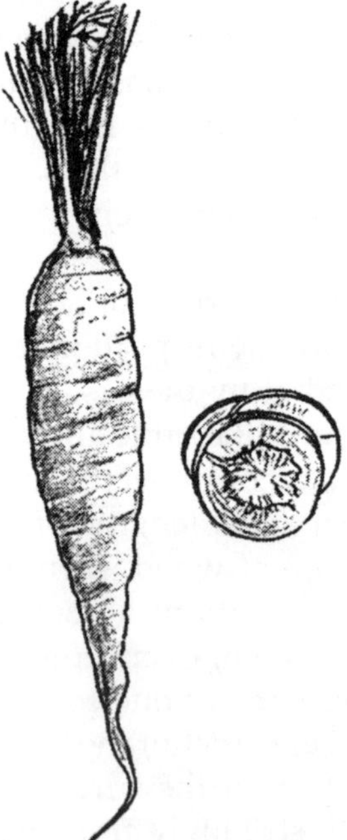

Fig. The Roots and the Stem are Part

As the plant becomes older, the root increases in thickness by the activity of cells between xylem and phloem, as stems do. This zone is at first a sinuous band rather than a smooth cylinder or square. Its activity is such, however, that it rounds itself out, forming the familiar central shaft of wood surrounded by a cylinder of phloem. Meanwhile the parts outside the central cylinder largely die and are lost; the epidermis and its hairs, the thin-walled, food storing cells beneath them—these usually form no part of the mature root. Corky cells appear around or in the phloem and form a surface which is hard and impervious to water, if the root becomes sufficiently old.

We are very familiar with some roots-on our plates at dinner: carrots, beets, turnips. The same is true of these special roots as of the general type described above: they are largely central cylinder, largely, in fact, wood. Their wood, however, is in a sense not fully matured —or has matured differently from most wood; for it consists of thin-walled cells which often contain some food, like the lost cells of the outer parts of the root. The many rings of a beet are not annual rings like those in a tree-trunk.

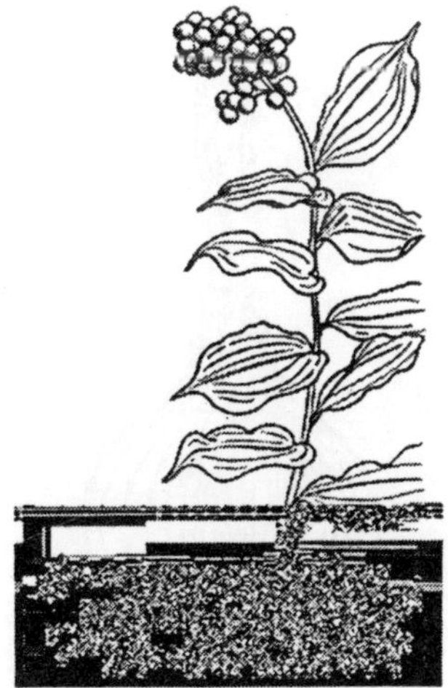

Fig. Grubbed Roots

They are zones of growth which appear one outside the other, each adding to the girth of the root like the original zone of growth which appeared between xylem and phloem. In a carrot you may see plainly the distinction between the wood, xylem and the surrounding bark, phloem; the wood is the central core, the bark is everything outside this. Both these parts rather fail to perform the functions usually characteristic of them and are useful to the plant—and to us —only as they contain an accumulation of foods, vitamins and other substances.

The American Indian lived mainly on the flesh that fell to his swift arrows, yet grubbed roots from the earth, using them sometimes for food, sometimes for medicine. Much curious botany has been lost with the spread of civilization over his forests and hunting grounds. Some of the wild fruits of the earth were bounteous.

The "man of the earth"—a relative of the sweet potato — has been known to reach a length of 4 feet and a weight of 20 pounds; starch enough for a family for some days. Primitive man all over the world founded his agriculture upon such gifts of nature, bringing the plant—or its seeds—into some small clearing more accessible to his lodge and thus assuring himself a meal in the future.

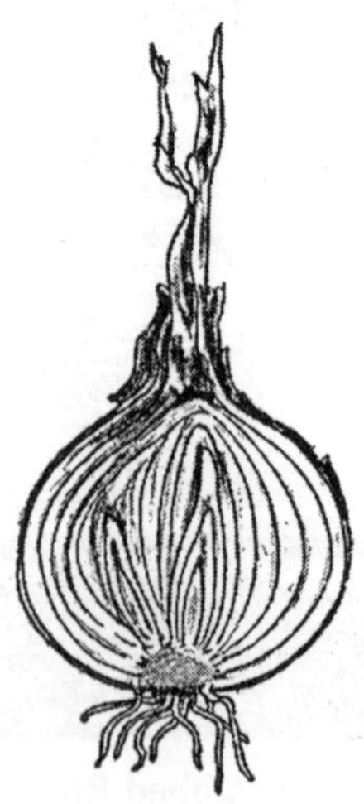

Fig. Earthy Crops

So civilization may be said to spring from the development by plants of swollen parts in which foods accumulated and became concentrated. Without these, herbivorous and crop-growing animals must needs graze like the deer and munch like cows; what sort of social culture could be based on such nutrition tasks the imagination.

The importance and prevalence of earthy crops is such that one is tempted to classify the parts of plants merely by the place where they are found.

It is quite generally supposed that underground parts of plants are roots. This is not true. When the Indians of the southern continent Learned to know the potato plant and dig up and roast its tubers, they were eating stems, as we now classify them.

This decision is based on their structure. A potato has a large central pith and conductive strands near the surface. Furthermore, it has buds-"eyes"—and the vestiges of leaves. It grows as a side-branch from the base of the main stem of the plant.

Underground stems are frequent. Many a wildflower of early spring pushes up from a stem which has lived safely beneath the surface all winter, filled with food. The red-juicy "root" of bloodroot (Sanguinaria) and the scaly underground part marked with the "Seal of Solomon" are known to many.

Bulbs also are classified by structure rather than use, place of growth, or even appearance; the botanical classification often intersects that of the horticulturist. The bulb of a dahlia is nothing more complex than a root, that of an iris is a stem; but a bulb of a tulip, hyacinth, or daffodil is a beautiful structure that merits a name of its own. (If one hesitates to dissect the germ of so much beauty, an onion will do as well.) These bulbs are really buds. Each has as its base a disc-shaped stem, from whose slightly convex upper side spring close-ranked leaves.

The outer leaves are scale-like, with hard and dry exterior surfaces. The inner ones are thick, fleshy, filled with food; some of their tips for a time extending into the air and light and becoming green, then wilting and withering away.

From the central point of the disc, which is the apex of the stem, finally springs the flowering branch.

Small bulbs arise in the axils of the leaves of the parent bulb and split it as they grow, so that in time it is replaced by a cluster of its offspring.

Other axillary branches may extend into the curious "droppers" of the tulip and some of its relatives, long hollow shoots which grow downwards into the earth bearing in their apices the germs of new bulbs.

Now if we consider in detail the work performed by stems and roots, the ways in which they serve the plants to which they belong, we may ask:

To what extent are the activities of the two organs the same and to what extent do they differ and do the resemblances and differences in architecture have any meaning in such a comparison? The answer is what might be expected; but the degree of adaptation that becomes evident exceeds the common idea of the fitness of plants for the world in which they live.

Great is the weight sustained by the trunk of a tree and incessant the attacks of wind and weather which it must meet; this is the virtue of that shaft of tough yet flexible wood. We sometimes wonder that a great tree may remain alive and thrifty even when its trunk is hollow, all the wood rotted out but a thin cylinder next the bark;—forgetting that it is *the material farthest from the centre* that is of the most use in resisting bending. A narrow strand of wire can be easily bent; take the same quantity of material and make it into a hollow pipe and it will be less easily distorted.

The wood in the centre is of little use in holding the tree upright in a wind; it is the wood near the bark that is of the greatest significance (in this as in the conduction of fluids).

Man had discovered and applied the same principles when he used hollow columns of iron to support a roof or a balcony.

The same principle also is still more vividly exemplified in a herbaceous stem. The centre of a geranium stem is merely pith, of no structural value; yet the stem is strict, stiffly upright, by virtue of the cylinder of woody strands near the epidermis.

A corn stem is held erect by its many single strands; these

are most thickly aggregated near the epidermis and there welded into a continuous cylinder by the aging of the cells around them; the fact that a mature corn stalk is hollow has little effect upon its pride.

Somewhat the same principle of mechanics is seen in the structural device known as an I-beam: a beam which is expanded into a thin web *in the plane in which the stress will come,* with most of the material in two solid strips to either side of the web; the function of the web is mostly to hold the solid strips apart and away from the centre of the beam. In a leaf the larger veins have cross sections like that of an I-beam; they extend in narrow vertical bands from epidermis to epidermis, holding these two fairly rigid layers apart.

The gigantic leaves of a banana, which tower from near the ground to twenty or thirty feet overhead, owe their stiffness to the same disposition of structural tissues.

Much of their interior is filled with air; long air channels, interrupted by numerous delicate septa, extend the length of the stalk and across the blade.

Separating the air chambers are narrow partitions which keep the two surfaces apart; the structurally important bundles of cells, to which the blade owes its support, are concentrated near the surfaces.

Fig. Morning Glory

The enormous leaves of the gigantic waterlily of tropical American rivers are strengthened by I-beams; their narrow ribs project vertically from the lower surface for several inches and are held erect only by other ribs which intersect them at right angles; these in turn are supported by buttresses that slope to the surface and have a function like that of the buttresses that strengthen the walls of medieval churches. Joseph Paxton, who designed the Crystal Palace erected in Hyde Park in 1851, claimed the leaf of Victoria regia as inspiration for the iron tracery which he created.

The flat expanses of leaves owe their stiffness to I-beams; the flexible yet unyielding stems which weather the violence of air, of rain and of snow and ice survive by virtue of the idea of the hollow column.

Both sorts of structure resist bending, both employ variants of the same underlying principle: removal of structural material from the centre. On the other hand a root is so made that—at least when young —it resists lateral stress very poorly; having its tough cells near the centre, it can be easily bent.

This would be disastrous—were the root exposed to such strains; actually it is embedded in dirt which supports it on all sides. The stresses to which it is subjected are mainly lengthwise; it is *pulled* by the efforts of the swaying stem to escape from earth. It is common knowledge that to obtain a structure which will resist pulls we weave strands into a compact cable so that the strength of the total is the combined strength of all the components, *even if some bending occurs.*

If we use the same number of strands separated from each other by an appreciable space, slight inequalities of stress, slight distortions or changes in the directions from which the pulls come, will act unevenly on the strands, subjecting now this one now that to a strain perhaps greater than it can singly sustain.

The strength of such a group is only that of any one strand by itself, as the strength of a chain is no greater than that of each link. A root whose supporting strands were scattered through its cross section would easily be broken; the roots that you find in nature, their vascular strands closely aligned in the centre, are as tough as ropes.

With biological complacency we extoll the adaptations of roots and stems and call them perfect since we could devise nothing better ourselves.

It is none the less remarkable that, in the course of their long history upon this earth, plants should have worked out structural reinforcements closely related to the different types of stresses to be sustained and effecting an evident economy in the use of the stiffening materials.

Without being too rashly explicit on the nature or plans—or indeed of the existence—of the Architect, it is thus that we may speak of the architecture of plants.

Chapter 4

Growth of Plants

A farmer has placed some dry grains in to the soil. One day the garden soil is smooth and hard. The next day it is seamed with small cracks and anon from every crack are thrust minute points of green. These take form as buds, variously folded and coiled and helmeted. Soon they shed the withered remnants of seed-coats and stand erect, offering diminutive leaves to the mild sun of spring.

We know their appearance is no accident, no miracle. They are in rows where the seeds were placed and they are of different kinds even as the seeds were of different kinds. Beans have at first two rather thick rounded leaves with a bud between. Peas have more delicate leaves, each compounded of several rounded leaflets and tipped with coiling tendrils. Squash first displays a pair of strap-shaped leaves, which spread apart like the leaves of a book as soon as they are pulled from the seed-coat. Morning-glories are the clowns of this seedling world, with three-fingered leaves which at first wear their seed-coats like dunce-caps.

Whence have they come, these vigorous infants, so busy, so sure of themselves and of destiny? What was in the dry seeds when, only a few days ago, they were put in the ground? According to the ancients the planted seed must first die; a miraculous resurrection occurs in the soil, the seed springs to life again and shapes itself in the image of its ancestors.

The physiologist, however, can demonstrate by using appropriate apparatus that the embryo, even in the unsoaked seed, is alive. It forms carbon dioxide. Respiration, that release of energy which accompanies the decomposition of food, continually occurs in it, perhaps at a rate so slow that it is difficult to detect. If many seeds are stored in a bin and particularly if they are not quite ripe and have not completely entered dormancy, the energy released in their respiration may partly appear as heat, which may result in a noticeable and even dangerous rise in temperature.

The rate at which their life goes depends largely upon water; when they are placed in moist soil the tempo is quickened; their long sleep is ended and they begin to grow. In much the same manner (though for different reasons) the bear and the woodchuck come forth with rising temperatures and quickening pulses after their winter of slumber.

Wonderful as is the awakening into activity of a dormant embryo, no miracle has occurred; no death and resurrection. Life is continuous from generation to generation. The activity of a plant pushing up from its soilcrack is but an exaggerated version of the activity of the inert but living seed which was recently planted, an activity which can be traced back to the time when it first took shape and individuality and became distinct from the parental tissues.

Fig. Under Leaf of Voctoria

It is clear also that in a seed the plant is already present as a formed individual. The parts which you see thrust up from the soil and those other parts which are not seen, growing in the contrary direction, are not suddenly formed from nothing, from an amorphous lump of life. Their origin may be traced back to the beginnings of the seed as a minute and delicate granule within the inmost cavity of a flower. In this rudiment of a seed the embryo began as a single cell.

The history of the origin of this cell and the manner of its formation belong in another chapter. We are here concerned with its fulfillment. In appearance it is unimpressive. No peculiarity distinguishes it from many a cell of stem or root; it has the ordinary parts of a cell and no more. There is no indication in its visible structure that in it lie the potentialities of a garden vegetable or a forest tree.

The outstanding virtue of this rudiment, this initial cell of a plant, is its really wonderful ability to divide itself into two. The details of this multiplication also are deferred to another place. Suffice it here that the process is a marvel of complexity and precision within that minute organism. The nucleus first divides itself into two nuclei; a new wall appears which makes a partition across the old cell cavity, one of the new nuclei being segregated into each compartment and the two half-size cells thus formed enlarge until each is the size of the original cell.

Even before this size is attained, however, they may divide again, each cell again forming two, which may again repeat the process and so on indefinitely. The result of this rapid geometric progression is a body of very small cells, actively dividing and enlarging, differing in appearance and in activity from any of the cell tissues hitherto described. This is embryonic tissue, Meristem. Meristic activity is the first symptom of growth.

As the first little nodule of cells becomes larger, some forces as yet quite unknown to our researches begin to shape it, to mold it into the form of a plant. By the time it becomes visible to ordinary means of observation, even before the seed is ripe and ready for harvesting, the embryo consists usually of a cylindrical stem, in most kinds of seeds split at one end into the slightly flattened rudiments of leaves. It increases in length more than in thickness and, perhaps compelled by the exigencies of space within the hardening seedcoats, may coil or double upon itself or become variously folded.

As the seed reaches maturity and is ready to be liberated from the parental envelops within which it has developed, the embryo ceases growth and practically suspends activity. It becomes the dormant germ which we find in a dry seed ready for planting.

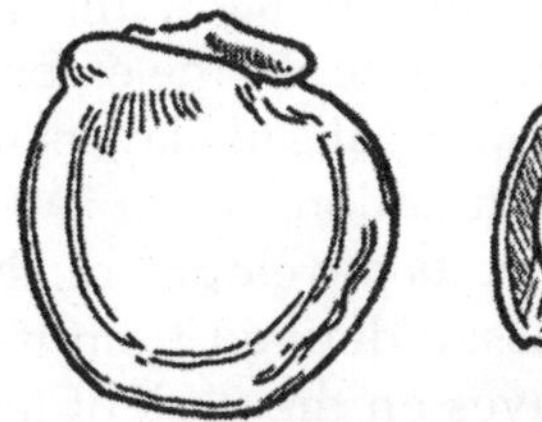 

Fig. Seed of Jimson Weed

Almost all textbooks of botany (those afflictions of the schools) begin with beans; their devotion to this vegetable has provoked derision. It is unfortunate that most seeds are too small for the unskilled hand to dissect; a bean is conveniently large. Other seeds will do as well, provided the means are at hand to study their minute parts; but a bean may (after a preliminary soaking in water) be easily dissected and comprehended without the aid even of a hand magnifier. So it is as well to join ourselves here to the vast throngs that have bowed themselves before this shrine of convenient knowledge. A bean is beautiful enough, whether we meet it on a plate or watch it break the soil in our garden.

If the skin or seed-coat of a soaked bean is slipped off, the interior part is revealed as formed of two separate halves; a pea or a peanut is made the same way. Separate—but not quite; at one side they are joined by fragile stalks to a small rod which lies along their edges. This is the stem mentioned above; the two parts joined to its apex are its two leaves—massive and unlike leaves though they be.

Break one off; there is disclosed, lying folded between them, the first bud at the extreme tip of the stem. On its other end microscopic study reveals the rudiment of the first root. Here, already formed in the dormant bean, is the new bean plant in miniature: leaves, stem, root. The animal (human or other) that munches beans tears apart with every mouthful a number of complete plants, each perfectly and beautifully formed, each capable of individual life.

The cells of this sleeping embryo, so lately moved by the common impulse of propagation, are largely uniform in shape and size. In their position and organization, however, there are traces of the future structure of stem and root and the transition region between these two. In the embryonic leaves the beginnings of epidermis and palisade layers may often be detected by microscopic examination.

During the early stages of its development, the embryo is embedded in a nutritious tissue derived from its parents, in which it feeds; so a worm lives on the flesh of the apple into which it has burrowed. In many seeds this food reservoir enlarges even more rapidly than the parasite within it and forms a conspicuous part of the ripe seed; to be finally used up during germination.

When we eat flour made from a buckwheat seed, we consume the food which otherwise would have nourished the seedling in its earliest growth; we take the embryo also, for good measure. In other kinds of seeds the embryo absorbs during its early growth—before the seed is ripe—all the food provided; greedily it takes in more than it can use at the time and comes to occupy with its distended leaves all the space within the seed-coat. Of this type is the bean embryo.

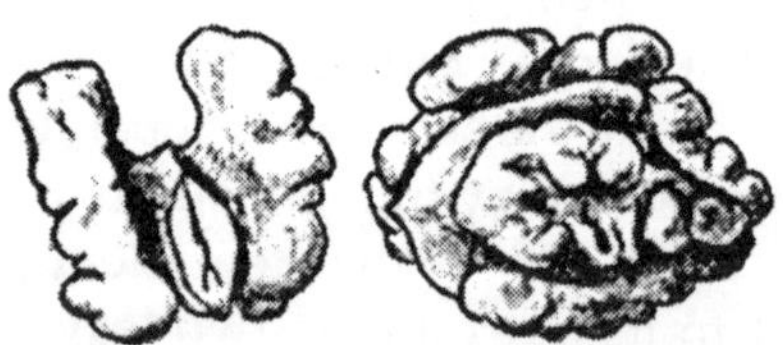

Fig. The Embryo

The two halves of the meat of such a seed, the two first leaves of the new plant, are swollen out of the semblance of leaves by the food with which they have gorged themselves. During germination they yield up the accumulated food to the actively growing regions of the embryo to which they belong.

The metamorphoses of seed-leaves are remarkable. In a walnut or pecan the wrinkled and lobed edible meat is really a pair of leaves; the sharp point at one end is the rudiment of a root; the stem tip lies buried in the overgrown leaves which grew from it, to be discovered by careful slicing. To eat beans or peanuts or peas or walnuts is to eat foliage. It seems a far cry from nuts and peas to lettuce and cabbage; but botanically, structurally, these are more closely related than almonds are to raisins or lettuce to radishes.

One large group of plants—sure to be encountered by the curious student or the assiduous amateur—does not conform with the general pattern of an embryo described above. In a lily seed, for instance, there is no sign of a pair of leaves at the end of the embryo; the same is true of a date seed or an onion seed and the seeds of a number of other familiar plants. When such an embryo germinates, the stem and leaves appear to originate from the *side* of the embryonic stem rather than from its tip.

Actually what is taken for the stem is mostly leaf, a usually rod-like leaf which terminates the stem; these plants have only *one* seed-leaf. The real tip of the stem on which this leaf grows is deflected to one side. The single seed-leaf may emerge from the seed, in germination, as a green needle. Or it may remain embedded in the food reservoir and act as a sort of food-absorber and digester for the rest of the embryo. From its base, at one side, springs the growing bud which rapidly sends up new leaves.

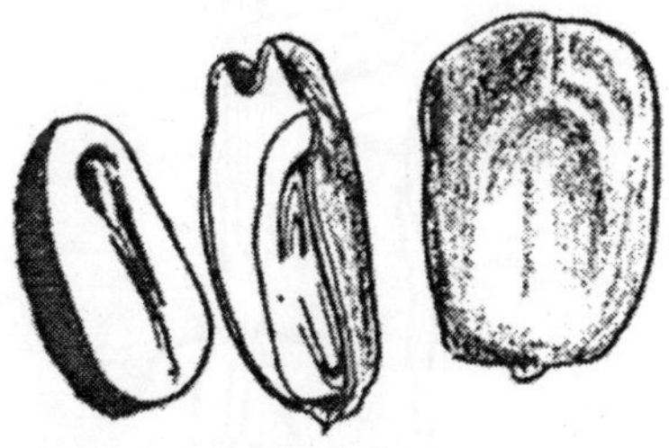

Fig. Corn Embryo

Perhaps most remarkable of all these one-leafed embryos is that of a grain of corn or wheat; embryos which we eat more often than any other kind, without a thought of their botanical peculiarity, but which also are involved in some of our economic processes in a peculiar way. Such an embryo has a curious leaf which embraces the rest of the embryo on three sides.

This object never acts like an ordinary leaf, never becomes

green, never reaches sun and air, never manufactures food. Nor does it contain a supply of accumulated food like the seed-leaves of a bean or pea. Its only usefulness to the embryo is as a digestive agent; it forms the digestive juices, the enzymes, which work on the starch which lies so abundantly against it and convert it into sugar which can pass (dissolved) into the cells of the embryo.

So to make malt we allow barley to germinate; in a short time it contains an excess of the valuable digestive substances —valuable to us as to the plant—and much of its starch has been converted into the more accessible sugar.

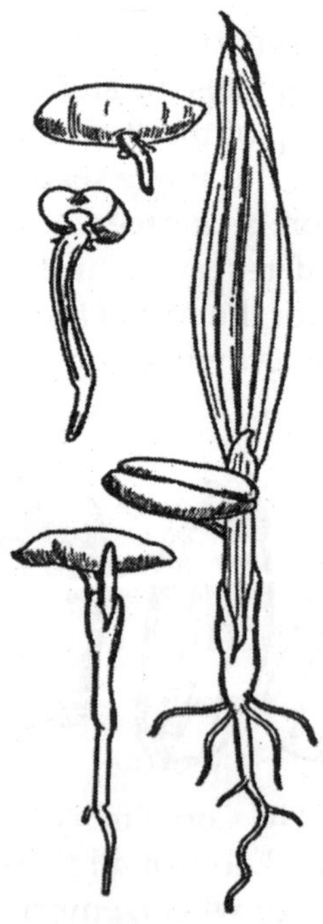

Fig. Growth of Seed

The plant has as yet only set the stage for its growth. The embryo is not much more than a small homogeneous mass of cells roughly shaped into the image of a plant. We plant our seeds and growth begins in earnest. Within a few weeks we have rows of plants which seem to have little relation to the minute and curiously shaped parts of the dormant seeds: tall stems, green leaves variously toothed and clothed with hair, perhaps tendrils; finally the buds which will be flowers.

It is apparent to even casual thought that the growth of any living thing, bean or baby, is more than mere increase in size. A bean plant is not a gigantic bean, nor is a man but an oversized infant. Not only does the grown plant have many more leaves than the embryo; these differ in shape and size from the seed-leaves. Branches of the stem and of the root have appeared and somehow acquired a complex structure.

The plant is no longer a body of similar cells; the cells have become specialized, differentiated, possessed of various features which limit their activities and confer upon them extraordinary powers. Such changes are responsible for the form of living plants (and animals) as we know them. All the marks by which we identify living things as cats and caterpillars and chrysanthemums and corn, all the arms, legs, eyes, ears, wings, hair, scales, feathers, leaves, roots, tendrils, spines, flowers, fruits: all these are formed during growth and their formation is an essential part of growth.

Growth is generally a combination of two phenomena: enlargement and maturation. Each may occur separately for a time. The embryo in its earliest development at first enlarges without much sign of organization into differentiated parts. The converse is also possible: a plant, or a part of a plant, may grow without becoming larger. Such changes—as in the colour and distribution of hair—are only too well known to human beings as they "grow older." But a general consideration of the growth of a plant or animal must involve both of these changes.

When a seed is planted, the reawakening of growth is evident first in the tip of the embryo opposite to the seedleaves: the tip which becomes the first root of the plant. As the seed-coats soften they are split and the white pointed tip of the root, an actively dividing and enlarging meristem, emerges through the opening it has made. It increases rapidly in length, turning down and pushing and wriggling its way among the particles of weathered rock and plant and animal remains which are soil. From it sprout thousands of minute transparent hairs, which penetrate the finest interstices of the soil, accommodating themselves to the individual shapes of the soil particles and following the films of water that surround them. These hairs increase many hundreds of times the area through which water may enter the plant. It has been calculated that the root hairs of a single grass plant, if placed end to end, would extend for *6,000 miles*.

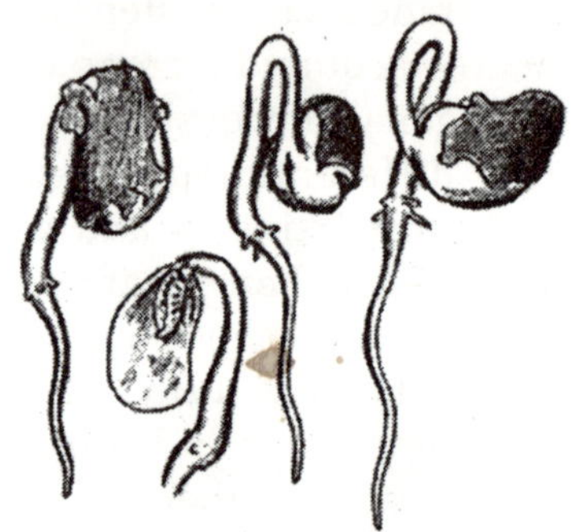

Fig. Seedlings

Without these the plant could hardly continue to move water up fast enough to replace its tremendous losses. Plant growers know that the absorptive regions are the tips of the roots and that the dirt must not be shaken from them, for this would destroy root hairs, the absorbing surface.

Now the stem of the embryo begins to grow. The bud which lies between the seed-leaves may grow straight up, forming new cells, new leaves, new lengths of stem, leaving the seed-leaves at its base to rot in the soil as they yield up their store of food. More commonly the stem beneath the seed leaves elongates; since one end has formed the primary root, now firmly anchored in the soil and the other end is attached to the seed-leaves still caught in the seed-coats,—since both ends are fixed, the lengthening stem becomes an arch, which breaks the surface of the soil as it were with its back.

Later the arched stem straightens, dragging its seed-leaves out backwards instead of pushing them ahead. The leaves spread apart, turning green also and exposing their flat surfaces to the sun; ready for the business of food-making.

Between them is the apex of the stem from which the stem continues its growth.

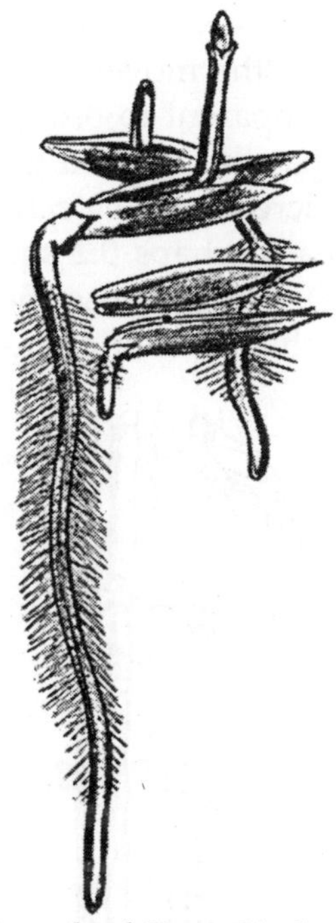

Fig. Growth of Grass Seed

All this, which has taken but a few minutes to read, occupies several hours or days.

The straightening of the stem and the expanding of the leaves are accomplished with curious waving motions; but these occur so slowly that to our eyes the plant seems to stand motionless.

Only when we see motion-pictures which have been exposed at a very slow rate during the period of germination and which are then projected at the usual rapid rate do we begin to have some conception of what is going on.

The young plant comes to life before our eyes. It grows a foot in a minute or two.

It sways to and fro, bows, waves its leaves, opens and closes them. Neighboring seedlings move in rhythm; the dance of life is not limited to animal creation. For full and vivid understanding of how a plant grows, we must watch it also through the microscope:—we must *see* it grow.

While growth, like the movement of the hour-hand of a clock, is usually not apparent to ordinary observation, if we arrange matters so that the tip of a growing root enters the field of vision of a microscope, we can readily see its motion across that small area (perhaps 0.2 square millimeter, about 3/10,000 square inch). We may use the root of a very small germinating seed, such as that of a grass.

Fig. Growing Root

This growing plantlet is so small that we can place it entire on the stage of the microscope in a drop of water and the lengthening root—its most active part—is so transparent that we can almost see every cell in it. These are the cells that are actually participating in the mechanical processes of growth;

in them originate the coordinated phenomena whose results we admire under the terms enlargement and differentiation.

Growth is not so rapid that we can watch, in a short time, many of the changes which occur in the living cell (though we *can* see them, with a little patience). But at a glance we can see cells in all phases of growth, which show what is happening as a series of "still" photographs can illustrate the flight of a bird or the muscular action of a running horse. The thimble-shaped cap which covers the extreme tip acts as a buffer and protector for the more delicate cells inside as it is pushed through the soil.

It is in the tissue immediately within the root cap that the processes of growth are initiated. Here the cells are very small; they are the most difficult of the cells of the root to distinguish. This part has retained the embryonic character; it is meristem, a region in which division of cells is active. Most of the cells, when they divide, form the new walls crosswise, so that the new cells are arranged in rows running the length of the root.

Next in order as our eye moves along the root, just above the meristem, is a region in which most of the cells are increasing greatly in size—principally in length. *These cells were meristem yesterday.* Even today it may be seen that some of them, particularly those nearest the tip, are still dividing, while their neighbors elongate. Elongation, which to a certain extent involves the cessation of division, sets in as a sort of wave which progresses down the embryonic root towards its apex.

Above the region of elongation we find another part in which enlargement has attained its maximum (a few cells may continue to divide). Here those differences between cells appear which result in their differentiation, their organization into tissues. Some on the surface sprout hairs, some form thick walls; some, which early concentrated on elongation and have attained greater length than their neighbors, undergo those complex changes which turn them into water-conducting vessels and food-conducting sievetubes.

These grass roots demonstrate also a very striking difference between plants and animals. The phases of growth

—elongation, maturation—progress in waves down the roots; the latter wave never quite catches up with the one ahead of it, so long as active life continues and neither of them quite catches up with the meristic productivity ahead of both.

The new cells formed by division in the meristem do not *all* elongate and mature; if they did, the growth of the root would be at that moment terminated, for maturation involves the cessation of division and there would be no remaining source of new cells (unless unusual stimuli were brought into play). Instead of this, the extreme apical cells of the meristic tip continue to divide, adding to the meristem, increasing its total number of cells; while the cells behind them cease to be meristem, enter upon elongation.

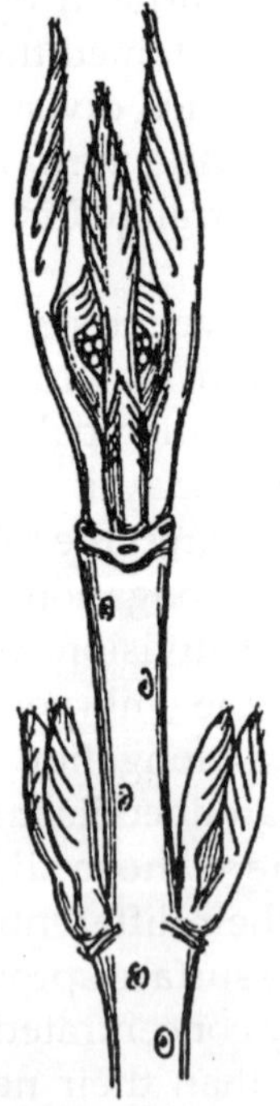

Fig. Growth of Stem Tip

This can be continued indefinitely. Meristem creates always new cells, some of which enlarge and mature while others repeat the division. The development of our own parts, our arms and legs, necessitated the formation of a certain number of cells, almost all of which reached their full stature and ceased division; there was no provision for continued addition of new segments, new joints, at the tip or elsewhere.

Perhaps it is better so; otherwise the air might be as full of elongated animal life as the soil is of roots. Only here and there in the animal body are there meristems which replace cells worn away or otherwise destroyed-cells of blood, bone, or skin.

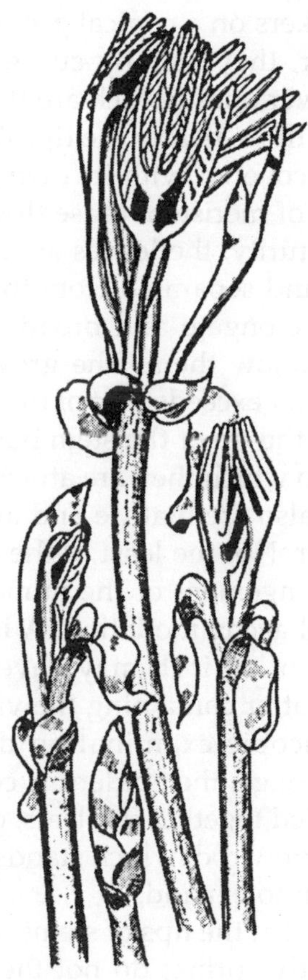

Fig. The whole Cluster is a Bud: Meristic Stem Covered with its Own Meristic Leaves.

In several respects the growth of stems is even more complex than that of roots. In addition to the rows of cells which form the tissues of the stem and which are comparable to those in a growing root, there are masses of cells dividing in other planes and projecting sidewise from the surface of the stem. These small bumps which so complicate the picture are the future leaves.

Leaves are formed from stem tips at the same time that the tissues of the stem are formed. At first the leaves are attached very close together over the surface of the meristem: slightly elevated puckers on a conical group of dividing cells. As they grow longer, they usually curve forward over the extreme apex, the lower ones (which are the oldest) being the longest and covering those nearer the tip. The whole cluster is a bud: meristic stem covered with its own meristic leaves. As the lower regions of meristem cease dividing and begin to lengthen out into maturity, the leaves attached to those parts are carried forward and separated from those behind. At the same time they also elongate and broaden and take on the shape by which we know them. The growth of their upper surfaces now somewhat exceeds that of the lower, so that they no longer curve over the tip of the stem but stand out from it.

When the stem to which they are attached has reached its maturity, the leaves also are mature and are at work making food and losing water. Not the least of the wonders of nature are the modes of arrangement of the miniature leaves in the bud and their mutual accommodation. A leaf no longer than the twenty-fifth part of an inch may have a distinguishable stalk and blade, the latter containing the visible rudiments of all its veins. Microscopic examination discloses even the palisade layer, the sponge, the epidermis containing stomata. The whole leaf is folded together in pleats, or variously rolled, forming usually a narrow rod, which stands with many others like it waiting its turn to expand.

Leaves are formed at the tips of stems. Consider again the dormant twigs of early spring; do not the leaves spring out along their sides in due order? Such is certainly their disposition, after a week or two of growth; but *they were formed last year*. In our temperate zone (a region which alternately experiences great extremes of temperature and moisture), the elongation of a twig occurs during only a few weeks in early summer; later changes are concerned only with its thickness and texture. It concludes its brief orgy of expansion by forming a new bud, usually at or near its tip; a bud which contains the apical meristem surrounded by the closely clustered meristic leaves.

The outermost of these at once mature, without greatly increasing in size, into hard and scale-like parts which effectively enclose the more tender young leaves and stem-tip within, sometimes being sealed together by a gummy exudation. In such condition the bud becomes dormant, awaits the following spring, nearly a year distant. When the new awakening comes, what a sudden urge to lengthen is there: the scale-leaves are burst asunder and usually fall; the leaves within enlarge to their full stature and, by the elongation of the stem on which they grow, are spread apart and disposed along its sides.

As the cells of leaves and stem mature, the regions at the base of the leaves in the angles between them and the stem-the armpits of the leaves—remain meristic, capable of growth, of the formation of new parts. Indeed, each of these small patches of cells usually becomes organized into a new bud itself, containing an apex which is a potential stem-tip, covered with the rudiments of leaves. These axillary buds usually become dormant when the terminal bud does so. The next spring, when the terminal bud pushes ahead with its programmeme, these likewise may grow out—into branches. Branches grow in a definite relation to leaves; beneath each branch is the scar of a leaf that has fallen.

Many lateral buds fail to awaken in the spring and remain seated above their leaf-scars; perhaps to be called into activity by some accident, some untoward event, in the parts of the branch above them. When a man prunes a fruit-tree, he makes his cut just above a point where a leaf was attached; the bud that lies there and its fellows for some distance down the twig, will break out into growth, fulfilling the beginning perhaps made many years before when that branch was first laid down. Wonderful is the vigorous surge of awakened life in the spring, which in a few days clothes the land anew with green. Like the rapid progress of a seedling, this is possible because the parts were ready in advance, formed last year, most delicately and beautifully folded and wrapped for a period of storage. The places where leaves will grow this spring were determined last summer; even the flowers (which, after all, are but a particular kind of branch) were then sketched in, sometimes in considerable detail and have now but to complete their organization and expand their calyces. In a few weeks now the leaves for *next* year will be made and the flowers; to lie inactive, waiting, through the long summer, through the fall and winter.

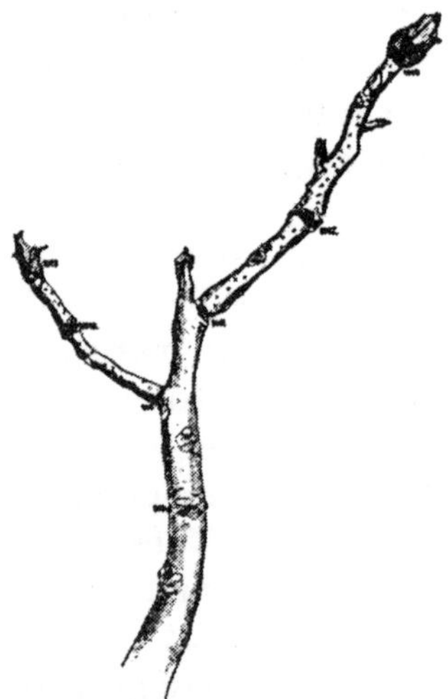

Fig. If the end of the Twig is Cut Off, these will Take up the Burden, Freed of the Dominance of the Tip

We may read the past of a branch by examining its branches and its scars. Every leaf worn during the last year left a characteristic mark when it fell—a scar by which the expert can identify the species. Just above each scar is a bud, which can perhaps grow into a branch this year (some trees have several buds in each axil). At the base of the length that was added last year is a ring of scars left by the falling of the bud-scales. Further down we can read the history of the year before last. We can still find the leafscars; above each is a branch, or perhaps a dormant bud that never opened.

Or above the leaf-scar may be another scar, formed by a flower cluster which matured into fruit and fell away. As each twig girds itself for the headlong course of spring, its buds compete for the water and nutrients sent up from below. The terminal bud usually has the best of it and grows the most; those nearest the apex often do nearly as well; further down we may find completely inhibited buds. If the end of the twig is cut off, these will take up the burden, freed of the dominance of the tip.

Each year new lengths are added *at the tips*; new lengths of every branch of stem and of root. Drive a nail in the stem anywhere; come back after many years; the nail has not changed its position, though the stem is much taller. Meanwhile that other growth described in the preceding chapter has transformed the twig, building new layers of wood and of bark, making it thicker, rougher, solider.

The meristems which accomplish increase in thickness work just like those at the apices except for the plane in which they divide and the direction in which they enlarge. Even within an old branch, within the trunk of an old tree, there are cells which have a capacity for division, for propagation, for the formation of new parts.

A stump will burgeon into new shoots; a felled redwood grows its circle of new trees from the still living roots; a redbud bole forms flowers as well as the young twigs. Living cells are constantly on guard within the seemingly inert shaft, ready to break out into healing layers of callus, to form new roots and buds if injury is done,—or in sheer exuberance to bloom.

Of the many kinds of men who go to and fro on the earth, we may notice two: those whose passion it is to describe things and those who inquire also into the causes. To search for causes is to attempt a special sort of description, for causes are those additional facts which immediately and invariably precede the facts observed. The attempt to *relate* phenomena in this way has resulted in the great mastery achieved by scientists over the physical universe—at least over portions of it. In the preceding pages we have attempted some description of the events and changes in a living plant which we sum up rather glibly under the term "growth." Though the description has been far from complete, enough has been said to convince the reader of the complexity of growth.

If we now ask what *causes* a living thing to grow, what factors control the particular directions that its growth takes, we may expect to consider an equally complex set of events and processes. In fact, the study of the factors of growth involves us at once in physics and chemistry and we shall probably not understand much of the actual causes of growth until we have attained to a greater mastery of these sciences.

The first discernible cause of the germination and growth of a seed seems, in its evident aspects, fairly simple: it is the entrance of water into the embryo. We have already said that a living plant is mostly water. Vegetation in general is about four-fifths water; a field of corn is so much water extending up from the soil into the air and held in position by thin membranes of protoplasm, cellulose and other materials. Water is the first need of a plant. The waters of the earth teem with life; the deserts are relatively lifeless. Since there is water on the planet Mars, it is legitimate to speculate on the possibility of life there; there is no life on arid Venus.

The dormant seed contains less water than an actively growing plant. The meristic cells of which the embryo is composed contain no large vacuoles full of water; even their protoplasm is less hydrated than more active protoplasm, the colloidal materials of which it is composed approach in their consistency a jelly rather than a fluid. This peculiarity may perhaps be connected with its dormancy. In any case, it springs to new life with the entrance of water which occurs when the seed is planted in moist soil.

Besides the water which becomes a part of the protoplasmic layer itself, water also accumulates in the vacuoles. Through a microscope we can actually watch the accumulation of free water in cells. A cell from the apical meristem of a grass root is almost homogeneous in aspect, relatively opaque. A millimeter or so farther up the root we can find a cell in which transparent spots have appeared; these are the vacuoles. As we watch they increase in size, touch each other, coalesce; what was protoplasm containing drops of water has become a drop of water surrounded by a thin film of protoplasm and perhaps penetrated by thin films of protoplasm.

The cell increases in size as the vacuoles increase and the one change is obviously related to the other. If we compare a young cell and a mature cell from the same root, it is evident that the latter has little if any more protoplasmic substance than the former. The same fact may be illustrated in another way. If we weigh some seeds, plant them and allow them to grow for several weeks, then dig them up, shake off the dirt and weigh the seedlings, we can record a great increase in weight.

If we now chop up the seedlings and place them in an oven in which all (or nearly all) the water in them will evaporate, then weigh the dried remains:—we find that the weight thus determined is little if any greater than that of the dormant seed. Indeed, if we grow our plants in the dark and determine accurately the weight of the dry matter of the seed and of the seedling, we find that the latter is *less* than the former. The seedlings actually lose dry matter in their growth, for by respiration some of the food originally present has been converted into gases which have passed out of the living tissues into the air; while the absence of light has precluded any possibility of the manufacture of food from carbon dioxide and water. The increase in size, several hundredfold, was almost entirely due to an increase in water content.

As the water rushes into the desiccated cells of the dormant embryo, into newly formed cells at the tips of roots and stems, it exerts tremendous pressures. An experiment familiar in classrooms, first performed by Stephen Hales in 1677, is to place some seeds in a jar of water and cover them with a lid to which is attached a weight.

The weight that the seeds will lift, as they take up water and swell, is astonishing. A jar of peas has been known to raise nearly 200 pounds. Students of anatomy long ago learned how to split skulls by placing peas in them and filling them with water. It is the existence of such forces within the growing embryo that enables it to force its way into the compact soil, sometimes pushing rocks aside or even fracturing them; the tender green shoots may lift paving-stones and masses of concrete.

If it seems strange that a liquid can exert such compulsion upon a solid, think for a moment of a fire-hose full of rapidly flowing water as compared with the same hose empty. It is sometimes asked: "How is it that such pressures do not burst the delicate cell walls and protoplasmic membranes within which they are exerted?" The answer is that sometimes they do; fruit and vegetables are occasionally split open during very wet weather. Normally, however, the millions of thin cell walls are supported on all sides by other cells pushing against them with equal pressures. It is always difficult to remember how small and how numerous the cells are.

What makes the water enter the seed? That story belongs to physics, for it is not peculiar to seeds or to living things. Blocks of wood and pieces of gelatin will swell when they are placed in water and often generate enormous pressures. A brief and very elementary account of some of the physical factors involved is here offered; the reader may pass it over if he does not care to pursue further the chain of causes.

It is known to the scientist, through a series of accurate experiments, that matter in general is composed of particles, very minute particles. At least he has proved that matter acts as if composed of particles and of late, using the electron microscope, he has even been able to make portraits of particles; though they are too extremely small to be visible to any human eye armed even with the most powerful lenses.

They are known as molecules. They, in turn, are composed of still smaller units called atoms and these are now known to be built of other units called electrons and their kind;—a hierarchy which dismays the imagination but offers no insuperable obstacle to analysis. It is with the molecules only that we are here concerned.

It is known further that molecules are regularly in motion or vibration, even in a body of matter which is apparently stationary and inert. In fact what we call heat is interpreted as due to molecular motion. The molecules of a gaseous substance move most rapidly, those of a liquid less rapidly, those of a solid with a speed that for most purposes is negligible. It is obvious, to draw a crude comparison, that the molecules must jostle each other like persons in a crowd; innumerable collisions constantly occur.

It is clear also that molecules at or near the margins of the crowd will not collide with their fellows so often as those in the centre and may indeed careen off into unoccupied regions. This is what happens when a gas "spreads" through a space; the phenomenon of diffusion can be admirably demonstrated by opening a bottle of ammonia. Any substance tends to diffuse, the direction of diffusion being normally *away from the region where it is most concentrated.*

So much is easy. What is not so readily understood is that each substance seems to act more or less independently of others so far as its motion is concerned, providing that no chemical interaction takes place. The vapour of peppermint oil will spread from a gum-chewing mouth through a room, in accordance with the principle just stated, in spite of the fact that air is already present in that space, air composed of molecules that are themselves in motion; in spite of clouds of tobacco-smoke moving in a contrary direction. Each substance free to diffuse moves *away from its own centre of concentration.*

We apply this principle to the living cell. Here we are concerned with liquids—everything that enters or leaves a living cell does so as a liquid. A young cell, like that of a dormant embryo, which contains relatively little water, is brought into contact with abundant water. The water moves rapidly and forcefully from the region of its greatest concentration *into* the cell, pushing aside molecules already there and probably entering into combination with some of them.

As the process continues, as water accumulates in the cell, the concentration *of water* within the cell membranes rises and the flow into the cell becomes less rapid. The cell, however, never contains pure water, but always a solution of various substances in water. Consequently a normal living cell surrounded by pure water is always distended to the utmost of its capacity by the water which diffuses into it.

We may easily reverse the movement. If living tissue—a piece of the stem or root of a plant, for instance a slice of potato or carrot—is surrounded not by water but by a solution of salt in water, water moves out from the living cells into the solution; for in this solution *water* is less concentrated than it is inside the cells. The cells shrink, the tissue becomes flabby instead of crisp.

If the process does not continue so far that other injuries to the cell set in, the flow may be again reversed, by immersing the flaccid piece of vegetable in pure water; in a short time it becomes as stiff as it was at first. Every housewife knows that lettuce and celery may thus be made crisp and juicy. It is known also that drinking a solution of salt or sugar induces thirst. That feeling is the conscious recognition of the loss of water by the cells of the lining of the alimentary tract.

It is easy to study with the microscope the loss of water by a living cell, to see these physical forces actually in operation. If a thin slice of plant tissue is mounted in a solution of salt instead of the usual water, the living contents shrink within their walls, leaving the latter unsupported. The cells—the living protoplasts—of a leaf become small green spheres which occupy only a fraction of the cavities within the cell walls.

The process of diffusion through membranes, commonly known as osmosis, accounts for much of what we observe of the behaviour of water in plants. It is thus that water passes from the outside to the inside of watery sacs, or in the opposite direction. But this is certainly not the whole story. Before it ever comes in contact with the living cells of the embryo, water penetrates the outer layers of the seed coat, some of which may be composed mainly of dead cell walls rather than of fluid protoplasm. The water finds its way between the particles and fibrils of the walls, combining with them in some way to form a colloidal structure, swollen and plastic. Dry gelatin will act in the same way, increasing in size many times and changing from a dry powder to a jellylike semi-solid. Certain gelatin-like plant substances, such as agar-agar, are sometimes used medicinally for this purpose. Administered dry, they absorb water and swell within the alimentary canal, forming a soft bulk which may have therapeutic value.

They are widely used also in the culture of bacteria and other small plants; the jelly holding in convenient form any solution with which we may desire to nourish them. A pinch of such material will bind in colloidal form fifty times its own mass of water, or more and hold it in a self-supporting, non-flowing mass which may line the sides of a test tube.

Even this process of imbibition does not complete the picture. It is suspected that other physical forces are set in motion by the activities of the protoplasm. It is an accurate statement that a cell is alive because it absorbs water; it may be said also that the cell absorbs water because it is alive. Whatever forces may be involved and however incomplete our understanding of the physics and chemistry necessary to explain them, it is clear that the rapid entrance of water into living cells and the development in them of tremendous pressures is a consequence of their material peculiarities and can be thought of in terms of physico-chemical laws rather than the mystic phrases of Hebrew prophets.

So large does water loom in our concept of growth that the reader may now fear that all we obtain by growing a crop is a rather intricately constructed reservoir of water. To correct this, it must be recalled that the solid matter of a plant becomes in time many hundred times as great as that of the seed from which it grew and may include the solid matter of many hundred new seeds like that. The plant that is grown in the dark cannot live indefinitely by gaining weight in water and losing it in solid content; after a few weeks of such hectic endeavor it, in fact, dies. The acquisition of dry matter is not so spectacular as that of water, not at first so rapid, but is just as essential to the ultimate increase of the plant.

The source of most of the dry matter of plants is already

known to the reader of these pages. It is made in the green parts of plants, from a gas taken from the air and from water taken from the soil. The first progeny of this union is the sugar glucose, which may or may not be converted into starch. In any case the surplus which accumulates in the leaves, if starch, is again converted into a sugar and this moves down through the stems and roots, in the vascular tissues.

Some of it reaches the growing parts of the plant, which make no food from inorganic matter, but which use a great quantity. Here part of it undergoes that process of decomposition known as respiration, is again converted into gas and water which forms no permanent part of the plant; energy being liberated in the process, energy which is used in motion, in constructive activity and in other ways. Of the remaining food—if any—some serves as the building material from which permanent parts of the cell are constructed.

The cellulose of the cell wall is a substance related chemically to sugar and to starch. With the intervention of certain enzymes produced by the protoplasm, by the aid of the energy released in respiration, the cell carries out this synthesis of sugar into cellulose, the new particles being deposited as microscopic fibrils at the boundary of the protoplasm and so adding to the thickness and the extent of the wall which encloses it.

The fats also and related substances, which form parts of the living protoplasm or which are deposited in the cell walls, are composed of the same atoms—carbon, hydrogen, oxygen—as is sugar and their synthesis begins with molecules of sugar which were sent down to the growing cells by the universal providers up in the sunshine. The most characteristic building-blocks of the protoplasm, however, are those substances known as proteins, which contain nitrogen in addition to the other kinds of atoms just mentioned and some of which contain also sulfur, phosphorus, or still other elements. These and many other kinds of atoms are obtained usually from the soil.

In recent years a new science—or perhaps a new art—has swum into the public view. Various names, fanciful, commercial and otherwise, have been bestowed upon it:

hydroponics, soilless culture, chemiculture, tank agriculture. The public is invited to believe that the beneficent persons whose business it is to sell "chemicals" have invented methods that will supersede the tillage of the soil.

The curse of Adam is finally to be lifted from us, the slow and painful labour of unenlightened man is to yield to the chemist's gentle arts. One needs only to dissolve in water the materials in a packet, following carefully the directions on the label, to have all that is requisite to the growing of successful crops in a small greenhouse or even in a parlor window. Several crops a year may be raised and it is said that the yield of each crop many times exceeds that obtained by the older culture.

Man always ready in the greed to believe that something can be had for nothing. True it is that all that a plant takes from the soil in which it stands can be purified, reduced to exceeding small compass and administered at will in solution. True it is also that more plants and a greater harvest can be raised in so many cubic feet of space than when one uses ordinary soil, ordinary temperatures, ordinary light. Yet everything that such plants take up from their purified solutions must come from soil, from some soil somewhere, by the sweat of brows. If the plants are to do more work, so also must their masters.

It is now many years that botanists have been growing plants, for experimental purposes, in water containing dissolved salts and other materials. As the science of chemistry developed, men did not delay to apply it to the study of living things and plant chemistry, biochemistry, also grew. Cultural experiments were the obvious means to learn what elements a plant takes from the world around it and in what proportions.

By such experiments botanists did indeed learn what all plants require for growth and how to supply it. Characteristically they ignored the possibilities of commercial application of their knowledge, concealing it in their own professional jargon; did not even pause to inquire whether such knowledge might prove a blessing or a curse to mankind.

Scientists advance in the confident belief that all knowledge is beautiful and blessed and so indeed it is—if we only know how to use it.

All plants take substantially the same chemical elements from their environment; some elements in relatively large amounts, some in very minute amounts. Carbon and oxygen we have already followed from the air into the leaves, which they enter together, joined in molecules of carbon dioxide. Oxygen penetrates the plant also through other portals, united with hydrogen to form the water which flows into the roots.

These three elements become, by the alchemy of green leaves, transformed into the carbohydrates: sugars, starches, celluloses and the like. Thus are a liquid and a gas metamorphosed into the solid matter of plants. Nitrogen gas is present in the air in large quantities; but unfortunately is unable to pass from the air into solution and into the cells of leaves.

Nitrogen, joined with other elements to make such soluble salts as the nitrates, is found also in soil; these salts enter the roots with the water in which they are dissolved. Such salts contain several elements joined together; thus do atoms of potassium, calcium, magnesium, sulfur, phosphorus, iron, boron, zinc, manganese, copper, molybdenum find their way into the plant.

The last five of these elements are usually present in extremely small quantities. Other elements also may be present in the soil solution and may enter the living tissues; but only those above named are known to be essential to the normal continuance of life. The absence of any one of these interrupts some normal activity of the plant in which it is involved and becomes rapidly noticeable in some abnormality of growth.

We do not yet know just how all these different elements are used in the plant, what parts they play. The importance of nitrogen has already been elucidated: it is used in the manufacture of proteins, which in turn form the basis of the protoplasm itself. There is a limit to the size attainable by a cell merely through the absorption of water; for continuing growth new protoplasm must be manufactured, new nuclei for new cells, new chloroplasts; this necessitates a supply of nitrogen. It is a tragic misfortune that man has learned to blast the bodies and works of his fellow men by means of explosives; many of these contain nitrogen, which is transformed into gas by the explosion and lost to the uses of life.

Sulfur also is used in the construction of plant proteins and phosphorus in certain other cellular constituents (found particularly in living nuclei). Sulfates, phosphates supply these needs from the soil. Calcium is used in the formation of the first thin layers upon which new cell walls are built up. So each of these elements has its particular share in the complicated industry of the making of new plant substance.

Those that come in from the soil are usually parts of inorganic compounds; they are assimilated into the structure of the living plant only if an organic building material is present also. Such a material is furnished by photosynthesis; all parts of the growing plant, all the protoplasm, proteins, fats and fat-like bodies, the waterproofing 'of the walls, the pectins which cement cells together, the walls themselves;—all begin with the organic molecules synthesized in green leaves under the influence of light; most depend also upon additional materials present in the soil.

It is sometimes said—it is a favourite catchword in elementary texts—that only chlorophyllous cells can make food. To make this a true statement it needs careful qualification. Only cells which possess chlorophyll can make organic food from purely inorganic materials. But every other cell in the living plant body—and in an animal body also —is endowed with the capacity to make foods if given at *least one organic material to start with*. Fats, proteins, the very substance of the protoplasm itself: all these are made in every cell that is alive; all are foods, at least for some organism.

In a world which is sometimes characterized as "running down," losing energy, losing structure,—in such a world the synthetic powers of protoplasm are its most distinctive feature. Throughout the living tissues of yonder tree, of every delicate herb in the garden, of your own familiar body:—manufacture is constantly in process, complex substances are continually being made from simpler ones, operations are being performed which are still the envy of the laboratory, wonderful as are its technical attainments.

The advertising of "chemiculture" depends for its success partly upon the mystification that has become associated with the word "chemicals." But indeed all things are chemical. Sticks and stones, sealing-wax and cabbages and kings, —all are composed of chemical atoms, associated in more or less complex fashions. Water is a "chemical," air is a mixture of "chemicals." The "salts" used in tank culture or in any other means of fertilizing plants have much in common with ordinary table salt, which also is a "chemical."

The substances used in the nutrition of plants have long been applied to the soil in manure or fertilizer. The former consists largely of the partly digested remains of plants, which must be decomposed before they are useful to another generation; the latter is composed usually of salts, the same precious chemicals put up into packages by the ingenious merchant and then sold by another name and at greatly increased prices.

The number of salts required is not large, for almost every one will contain more than one of the few needful elements; potassium nitrate contains potassium and nitrogen, magnesium sulfate magnesium and sulfur. A satisfactory solution for the culture of plants may be made by dissolving in water the appropriate amounts of potassium phosphate, potassium nitrate, calcium nitrate and magnesium sulfate.

The other elements, required only in minute quantities, will quite probably be present as "impurities" in these salts. The fertilizers applied to garden soils contain much the same substances. The correct quantities and other necessary particulars, can be obtained from many works on gardening and on plant growth; certain of the agricultural experiment stations have issued bulletins on these subjects.

Enough has perhaps been said to make it clear that the growth of plants does not depend solely upon the chemical elements which it can secure through its roots. The use of these elements is dependent upon the presence of organic substances made by the plant; the manufacture of these in turn is dependent upon sunlight (among other factors). That is, it is of no use to supply plants with the most scientifically accurate nutrient solutions unless the light is sufficient for a corresponding supply of organic food.

This is the "catch" in the praiseworthy project to produce several crops a year. Plants do not add to their substance during the night and little if any in the weak light of winter days. Natural light may in truth be supplemented (even replaced) by artificial light; but this is expensive, eats up the profits of the crop.

In general persons inexperienced in the growing of plants and in chemical technique have been disappointed and others who believe too readily the statements on the packages are likely to be disappointed. It has not yet been demonstrated that more and better plants can be raised without soil than with fertile soil. The undeniable advantages of tank culture are that large crops may be concentrated in small spaces and conveniently near to large centers of population. But the same substances must be supplied to the plants as if they had grown more widely spaced and at some distance from the point of consumption and the proper administration of the proper substances is likely to involve considerable capital and more than a little operating experience and skill.

So, finally, we have seen our plants emerge from cracks in the soil, unfurl their different parts, absorb water from the soil, exhaust their inheritance of food and begin to acquire and manufacture their own supplies. In all this long and sometimes tedious account we have carefully omitted to call attention to the most obvious and in some ways the most remarkable aspect of growth. One group of phenomena has yielded, quite recently, such notable results in our attempts to understand and to control life that it is here offered as a succulent dessert after the solid fare so far furnished.

The most remarkable facts of nature are so obvious that they escape our attention. Living creatures are adjusted with extreme nicety to the demands of particular environments and the continuance of their lives depends upon this adjustment.

When the farmer plants his seeds, they roll into their appointed places and come to rest in all sorts of positions.

Within the seed coat lies the dormant embryo, from one end of which the root will spring; this seed lies with its rudiment of a root aimed skyward, that one directs it towards the centre of the earth, the majority point it at random in all other possible directions. Yet when germination occurs, when the root-tip breaks the seed-coat, it grows *down* into the soil. If not already directed downward, it bends in its growth, quickly, unerringly, so that its further extension is downwards.

"But of course! What could be more natural?" Precisely: this is a *natural* phenomenon, which is to say that it has a *cause*. What makes the root grow downwards? And, while we are on the subject, what makes the stem grow upwards? Both actions are equally natural. Certain types of minds are content to answer that roots grow down because it is necessary that they do; if they grew upwards the results would certainly be disastrous.

The essential sterility of such a point of view is discussed elsewhere in this book; all that is necessary here is to point out that it does not answer our question—it does not give us a *cause*. We are asking, not what advantage accrues to the plant by the downturning of the root, not what would be the consequences of its failure to do so, but what factor, or set of factors, in the root or in its environment, acts upon it to *make* it grow down. The strictly natural elements of the situation are what we are after, not any possible *supernatural* aspects.

Until quite recently science could furnish no answer to such questions. Charles Darwin discovered, about 1880, that a decapitated root-tip, one from which the root-cap had been removed, did not respond in the usual way; it seemed not to know which way was down. He showed also that the bending of a stem towards light depends on its tip.

From this clue scientists plunged into grave speculations on the "perception" of direction by a root- or stem-tip, the "transmission" of an impulse to the actively growing cells behind the tip and the "response" of the region of elongation which effects the actual bending. But not until 1910 did further experiments shatter their vague theories and pompous terminology and open the way to actual knowledge of the factors involved.

Like most decisive experiments, those which illuminated this subject were very simple; so simple that everyone wondered why *he* had not thought of them. A Danish investigator, Peter Boysen Jensen, cut off the tips of oat sprouts and then stuck them back on with gelatin. Such tips, when illuminated from one side, caused the lower part, below the gelatin, to bend just as oat sprouts normally bend towards light. Whatever the "stimulus" which travels down from the tip, it can pass through gelatin.

Could it be a substance, a diffusible substance dissolved in water? Boysen Jensen and other researchers, in a beautiful series of experiments, showed that what travels through the gelatin is indeed a moving substance, a chemical substance which can be isolated, removed from the plant, studied by it. It can be collected by cutting off the growing tips of oats or other plants and placing them on minute blocks of gelatin or agar; the substance formed in the tip diffuses out (doubtless dissolved in water) and into the jelly.

The blocks are then empowered to work miracles. Perch one on one edge of a decapitated shoot and the shoot will bend *just as if light were coming from one side.* The direction of the bending can be controlled at will. The cause of the bending becomes clear when we realise that the diffusible substance that entered the gelatin from the severed tip can now move down into the shoot. If it goes down equally on all sides, growth is equal on all sides and there is no bending. If it goes down one side more than another, *growth is stimulated* on that side more than on the other; one side of the shoot grows faster than the other and the shoot becomes curved.

From such early simple experiments has arisen a great body of exact scientific knowledge and a highly developed technique of investigation. The substances which the early workers trapped in their little blocks of gelatin are now included among a number of materials rather loosely known as "growth substances." Several have been isolated chemically from growing tips of roots and shoots; these are usually known as auxins.

Their action may be duplicated by certain complex organic substances not normally found in plant tissues. By using such "hetero-auxins" various curvatures and other types of abnormal growths may be obtained and much new information about the processes and factors of growth is being obtained. Such substances have also found their way on to the market and, properly used, may help to stimulate the more rapid production of roots by cuttings. The explanation of the "tropisms" of plants—thus are such curvatures of growth designated—is not complete; many points remain to be investigated.

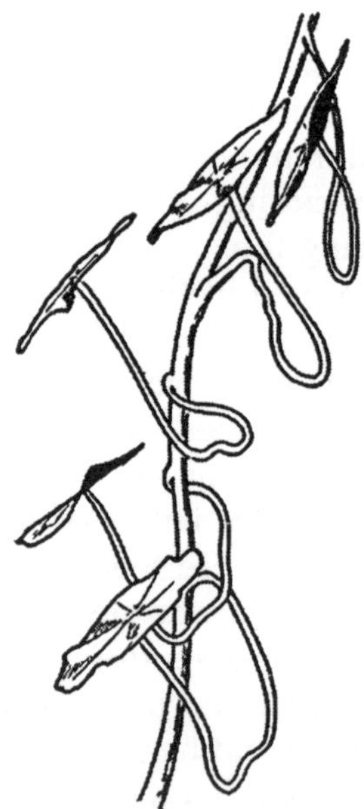

Fig. Growth Towards Light

But experimental science has shown that we may think of these "natural" responses to the environment in terms of actual substances, substances which may be extracted from plants and studied chemically, substances which are not by themselves alive nor in any essential way different from substances which occur in inanimate nature, although they are normally manufactured by living cells. Apparently the auxins are made by the meristems at the tips of growing roots and shoots (and perhaps elsewhere in the plant) and diffuse away from these apices into the older portions.

If light is more intense on one side of the growing shoot than on the other, more auxin is supplied to the shaded side (the reasons for this movement being as yet unknown) and stimulate that side to greater growth.

If the stem is in a horizontal position, the lower side seems to receive more auxin and consequently grows more rapidly, so that the tip curves upwards. In a root on the other hand, the accumulation of an excess of auxin on the lower side seems to retard development on that side, so that the tip curves downwards; root cells are apparently more sensitive to auxins than are stem cells (as some persons are more affected by alcohol or caffeine than others) and are affected adversely by concentrations which merely act as a pleasant stimulant on stems.

Teachers have been telling their students for many years that roots grow down and stems grow up "in response to gravity." This is true; but it made no sense to most of the students. How could gravity make something go upwards?

And above all how could a force in one direction result at the same time in two actions differing in direction? The truth may now be stated a little more intelligibly. If two unequal weights are attached to a rope which runs over a pulley, the greater weight will descend and the smaller one rise; *both actions are due to gravity,* directly or indirectly.

All that is necessary is the intermediacy of the rope and the pulley. So, if we understand something of the actions and movements of the auxins, we can understand the effect of gravity on plant growth. To put it crudely, gravity causes the auxins to accumulate on the lower sides of horizontal stems.

There they interact with other constituents of the cells in such a manner as to influence these cells to elongate more rapidly and to a greater extent; this forces the growing tip of the stem upwards. The auxins are the large weight, the stem cells the rope and pulley and the tip the small weight. Gravity does not make something "fall upwards"; that indeed is a contradiction in terms. But gravity can set in motion machinery which will lift something. So also a man rowing a boat across a stream may respond to the drag of the current so as to keep the boat's nose headed upstream, against the current.

It was no idle nor facetious mood that prompted the saying that a decapitated root seemed not to "know which way was down." It seems a bit strained to speak of a root as "knowing" or "not knowing" anything. Knowledge is usually associated exclusively with animal life. But careful consideration of certain aspects of animal knowing may disturb the satisfaction with which we view the alleged superiority of the animal realm. For an animal to *know* something about the world around him he must receive certain stimuli, which effect certain changes in certain of his parts.

The fact that knowledge of "down" in an animal is not essentially different from the mechanism which governs the direction of growth of a root has been rather amusingly shown by experiments performed upon crayfish. These crustacea possess small cavities through which water flows and in which sand grains usually accumulate.

The animals subjected to experiment were denied sand and instead furnished with iron filings. When they were placed in the field of a magnet, they acted as if the magnet were "down," swimming on their sides or even on their backs with complete disregard of the absurdity of the situation.

Their "knowledge" of direction was ordered by the position of the iron particles inside their bodies; the direction in which these moved was "down" to them and they moved so as to get these against the side of the internal cavity usually nearest the centre of gravity. Thus do these animals normally remain "right side up" in a difficult world by a simple mechanism which has some analogies with the tropistic mechanism of a stem or root.

One of the most brilliant chapters of modern biology is the investigation of the so-called vitamins. Though this story seems to concern animal life—and indeed human health — more than the plant life which is our present subject, it has recently become clear that the vitamins are as essential for plants as for animals.

Our knowledge of this subject is still growing and growing with remarkable velocity; it begins to appear that we are on the threshold of being able to make some sweeping generalizations about the relations between life and certain classes of substances, in which the vitamins and the growth substances are included.

Throughout this book life—the total activity of living protoplasm—is presented in terms of physical and chemical processes. This point of view is enhanced by history: the great discoveries of the past hundred years in biology have been made largely by men trained in physics and chemistry as well as in botany and zoology. Nutrition, as we have seen, is to be thought of in terms of specific elements and compounds. But when scientists reversed the process of analysis and tried to synthesize chemically pure foods from these elements and compounds, they soon found that something was lacking. The animals fed on such foods died.

They soon discovered also that a lot of animals who were becoming very sick on a supposedly complete and chemically pure diet could be rapidly restored to health by the addition of some natural food to the diet: egg-yolk, milk, cod-liver oil were some of the foods which supplied the deficiency. When we consider the list of such "accessory foods" we find it difficult to discover anything that they have in common; certainly none of them has any chemical element that is lacking in the so inadequate synthetic diets. The clue to the solution of the problem was the realization that these accessory foods were also those which were necessary in human diet to prevent certain diseases.

The dreadful affliction called scurvy had been known for centuries and was an all too familiar scourge of men at sea for long periods and of the poor during winter. The surprising discovery that it could be absolutely prevented by adding oranges or lemons to the diet made Captain Cook's famous voyage the beginning of a new epoch in human living. The terrible oriental disease called beri-beri was extirpated from the Japanese navy by the addition to the diet of certain things removed from rice by the modern milling.

The pitiful malformations known as rickets were prevalent in countries and during seasons with little sunshine and could be cured by exposure to the sun—or by taking cod-liver oil. The common element in these "deficiency diseases" is that any of them can be cured with extraordinary rapidity by the addition of minute quantities of certain foods to the diet. In 1912 Casimir Funk proposed his famous hypothesis that all these dietary necessities contained substances of the chemical class named amines, substances necessary to life: lifeamines, vitamines. The name stuck. We know now that they are not all amines and spell it vitamins.

Only recently has it been discovered that plants also use vitamins. Roots just emerging from a seed can be cut off and grown by themselves in a flask of nutrient solution, provided certain vitamins are present in the solution. Before this was realized, attempts to grow isolated roots failed; since it has been known, laboratories have become cluttered with large root systems which grow in flasks without any tops at all; scientifically valuable if agriculturally impotent.

In nature the roots derive their supply of vitamins from the same source as their organic foods: from the tops. In general, green plants manufacture vitamins sufficient for their own needs; a fact which we utilize in supplying ourselves with these vital necessities. They do not need vitamins in the same sense as we do—they need not be supplied with them from the outside. But they do need them in the sense that the living protoplasm of their cells fails in its activities unless the proper vitamins are present in the cells.

Evidence is increasing also that plants and animals use the same vitamins in the same way—in the *same processes fundamental to all life.* It is astonishing how in the midst of the diversity of the living world we are brought back again and again to the unity of life. You, O sovereign and intelligent man, master of the earth, must consume your daily stint of thiamine for the same reason as the mold on a bit of garbage must have it; you and that lowly fungus are closely related in some primary physiological pattern. The molds, mushrooms and those penetrating parasites that cause diseases of other plants, are classed together as the Fungi; they all lack chlorophyll. Consequently they are unable to manufacture food by photosynthesis and must obtain organic substances (as we must) from their environment. Many of them must obtain vitamins also from the materials in which they grow. Some species can make all their vitamins, some can make some of their vitamins and some can make none of those with which we have become acquainted.

Fungi may be easily grown in small tubes called test tubes, or in flasks, on substrata which contain exactly measured quantities of known and chemically purified ingredients. The vitamins (many of which may now be obtained in chemically pure form, their exact nature being known) may likewise be administered, not by squeezing an orange into the culture medium, but by adding definite quantities of carefully prepared solutions.

The minuteness of the quantities necessary to sustain life and to permit growth is fantastic, almost incredible. Recently I saw an exhibit of two test tubes containing a solution of various foods and other substances into which a fungus had been introduced. In one tube an abundant growth covered the surface of the food; in the other—nothing. The fungus in both was the same, the time and temperature and light to which the two cultures had been subjected were the same, the food was the same-except that in the first a vitamin had been placed, in the second it was lacking.

This particular vitamin may be purchased in pure form at a cost of $10.00 for 0.075 mg., seventyfive ten-thousandths of a milligram or about two-and-onehalf millionths of a millionth of an ounce. The price amounts to something over $62,000,000 a pound. And, at this rate, the quantity of the vitamin necessary to differentiate between the two test tubes, to permit abundant growth in the one, was about *one cent's worth!*

The effects of some vitamins are detectable when even smaller quantities than this are present. Inexpressibly minute as the trillionth of an ounce seems to us, it is still a fairly large amount for a chemist. Atoms and molecules are much smaller. The pennyworth of vitamin mentioned above contained many *millions* of molecules. As human beings we suffer from the deficiencies of our sense organs.

We are so large, so coarse, that we cannot perceive even cells without the aid of a microscope, far less molecules and atoms and we have trouble in imagining the teeming and active world of units which are invisible to us. At the other extreme we are so small that we have equal trouble in extending our mental tentacles enough to grasp the stars and the facts of stellar space.

Biologists, philosophers and others,—a motley crew-have prattled for centuries about the "mystery of life," and have sought to specify the single peculiarity which distinguishes living from non-living matter. On this attitude more in the next chapter; it is pertinent here to notice that one of the most striking and peculiar characteristics of living matter is one that has been scarcely considered as a candidate for divine honors. This is its relation to substances present in extremely minute quantities.

Fig. The Bean that Sprouts from the Cracked Soil

Every natural process to whose core we manage to penetrate is likely to reveal itself as conditioned by some enzyme, some vitamin, some hormone, or some system of such substances. They are in no sense mysterious—as the popular press loves to label them. They are definite chemical substances, many obtainable (at a cost) in pure form. Even their mode of action is yielding to careful physiological research and what we do not know today will be disclosed by the science of tomorrow. Certainly this fact, that so much of the life of protoplasm is absolutely dependent upon fantastically small quantities of certain organic substances, is one of its most distinctive properties. The bean that sprouts from the cracked soil of your garden is digesting its insoluble foods by means of enzymes which its protoplasm makes, it is righting itself in the world by means of auxins which its apical meristems make, its roots and its stem are elongating by virtue of vitamins which the green parts make.

These substances, these and others like them, are what ensures that the bean plant shall behave like a bean plant, shall be a coordinated mechanism, an organism, rather than a mere sprawling amorphous lump of expanding matter.

Chapter 5

Purposes of Plants

Outside the window the oriole is calling in sweet full tones. The shriller melody of a song sparrow, oft repeated, sounds from across the garden. My thought is invaded by the scent of new-mown grass and apple blossoms are bursting their buds. A fly buzzes against the glass of the window and a very small spider sets up housekeeping in a corner of the floor. My own fingers move like mad dancers over the keys of the typewriter.

Further afield, in all directions, for thousands of miles, countless living creatures are waking into renewed life. The prairies and woods, mountains and deserts, lakes and oceans and rivers teem with living things, which creep or gallop or slink or swim or fly or thrust down roots in their innumerable versions of life.

What is life? What principle, what fluid or force or urge pervades certain assemblages of atoms, certain structural units and informs their behaviour? Rivers of ink have been poured out in attempts to answer such questions. Biologists and philosophers, each in turn misleading the other, have tramped the ground into a morass upon which it profits not to venture. He who lacks philosophical curiosities may well skirt this dubious region and pass directly to subsequent chapters.

Yet it is possible, without attempting to explore the swamp, to point out certain dry spots, to notice certain landmarks, to make certain distinctions. If the reader has indeed a desire to evaluate the science of living things, to penetrate with the clear light of reason the curious murk which has gathered about its central statements:—then let him undertake a brief excursion on a philosophical path. What is life? The difficulty with this question is not wholly the fault of the scientist; it is partly that of the question. In a sense this is an improper question to ask of science. Science, having endured periods of contempt and persecution, now consorts with royalty and plutocrats. Everywhere the laboratories hum with activity; printing presses turn out their monthly acres of research. Yet it is not easy to explain what science is, how it differs from other human activities. It has been rather widely misrepresented, even by those who purport to teach it. Like many another honorable name, the word has been a frequent cloak for bigotry, fraud and predatory commerce.

It is written in a newspaper that "science has not yet perfected a train window that can be opened..." but that "fighting flames in skyscrapers has developed into a science in New York City."

These statements are symptoms of the most common view. Science is machinery, shining metal, turning wheels, automatic control. There is evident confusion here between science and invention. Invention is as old as man.

The man who first fashioned a rude vehicle to go upon round wooden discs, the man who first used the elasticity of bent wood in securing his food,—these were greater inventors than Edison, nameless and unsung though they be. Such accomplishments, whether lost in the years before history began, or recorded on the bronze tablets of a self-conscious civilization, display a creative imagination which is perhaps the most encouraging feature of man's stay upon earth. But this is not science.

Science, *Scientia,* means knowledge. Any knowledge has at least two roots. One—perhaps the most important—is an idea or a system of ideas, a theory about things; the other is the coincidence of ideas and experience. The latter is called EVIDENCE and the characteristic feature of modern science, that which accounts for most of its success, is its insistence upon evidence. The ideas of science must be amenable to testing by experience.

If you are curious about the nature of science, you must

go into the laboratory and see what is done there. You will find men and women weighing things with extremely delicate and beautiful balances. You will find them peering through microscopes, spectroscopes, telescopes. You will find them watching needles which waver across graduated dials. The recondite facts of worlds within worlds, the play of electrons and the energies of stars,—all are translated by appropriate instruments into lines, dots, numbers, marks on a photographic plate, simple symbols *upon which a number of trained observers can agree.*

Thus are "facts" obtained, facts selected by the light of ideas. This is the "objectiveness" of science: the findings are accessible to all and can be duplicated (within certain limits called experimental). The superiority enjoyed by science over magic is simply that *it works.* An object does weigh the same again and again, two substances whenever they are brought together continue to yield the same new compound, the stars and electrons do move in their courses; nature, *in these terms,* becomes predictable.

In spite of the naive bragging of scientists, it is probable that scientific method does not represent the ultimate human activity. Much of human experience is inaccessible to it. The physicist can readily enough translate for you into numbers the identity of a sensation which you call pink; but in the process he loses the pinkness of it. You can measure the frequency of vibrations of light reflected from the petals of a rose; but you cannot explain in any terms which have the same meaning to another human being how "pink" feels to you.

The peculiar validity of science rests on the kind of evidence which it demands and this also limits its scope. So if we adopt as our sole method of transportation an automobile, by its very excellence we are debarred the reaches of the upper air and the solitary wastes of ocean.

Is it not now evident that to ask of a biologist what is life is to suggest that he be untrue to his method? "Life" is what we live; the function of that inmost ego that can hardly hold a mirror to its own features. Though the most commonplace fact of existence, it is confusingly veiled in our likes, dislikes, passions, purposes, all the flow of experience which makes a private being. Possibly indeed it is no more than a name for the personality born out of these conflicts. It is doubtless because I think of myself as *one* person that I think of life as *one* function and with this in mind I shape my question for the biologist. It sticks in our mind that *this* is life and *this* is what we ask the scientists to elucidate; forgetting that such a matter may not be translated into the readings of his instruments.

The assumption that "life" refers to a single function is deep-rooted in the history of science. Two hundred years ago Linnaeus directed the attention of scientists to the classes of nature. *Lapides crescunt,* he wrote. *Vegetabilia crescunt et vivunt. Animalia crescunt, vivunt et sentiunt.* Stones grow, plants grow and live, animals grow, live and feel. Eighty years later an eminent botanist had progressed so far as to translate the "mysterious matter" into physical terms and to contrast "vital force" with molecular attraction. The supposed single difference between living and dead emerges from its dim religious background and assumes the impressive disguise of scientific terminology.

Let it be granted that life is a mystery, the greatest of all mysteries. Let it be granted also that a discourse upon life may be a poem or a gospel and that truth may reside therein. Yet this is not what we ask of the naturalist when we demand that he expound the nature of life; nor can such discourse claim the prestige of scientific method for its own. If we expect a rational answer, let us recast the question about life. What are the characteristics of life discernible to the biologist and which of them are peculiar to the living condition? If there is indeed one difference—the essential rabbit—that also we shall learn, at the end of our analysis. There is no magic key to the caverns of knowledge, no Open Sesame; we must search for evidence.

An amusing toy sometimes called a "marine garden" is made by immersing certain crystals in appropriate solutions. A crystal of potassium ferrocyanide (for instance) in a solution of copper sulfate enlarges as you watch it, sending out twig-like protrusions which branch in a manner recalling some aquatic plants. The observer inevitably likens it to a living thing. It absorbs water by osmosis and the material upon which it "feeds" is changed to the material of which the "organism" is composed—features of growth which, in many textbooks, are falsely described as peculiar to living things. Linnaeus knew better.

The growth of a living creature is obviously much more complex than the growth of a crystal or a mountain or a cloud; but it seems (so far as present facts reach) no different *in principle*. The forces and substances involved are found also in inanimate nature, working in essentially the same way.

Such is the somewhat disconcerting conclusion that emerges from the tremendous wealth of biologic data now available: living structures, living processes, do not differ from non-living structures and processes in any way discernible to science save in the complexity of their organization. Chemical change, the synthesis, digestion and oxidation of organic substances, the absorption and manufacture of gases, the movements of molecules in osmosis;—all are closely comparable with events which occur outside any living cells.

That living things reproduce their kind is often adduced as their distinctive feature. In a popular encyclopedia for "boys and girls and their parents" we are informed that true organisms are distinguished by this capacity; we learn that one could imagine a living thing lacking the ability to reproduce, but that it simply does not exist. Alas the poor mule! It is evidently not a "true" organism. To such absurdities are educators (who doubtless know better) carried by the traditions and suppositions of the vitalistic approach.

The subject of reproduction is indeed complex; but no more so than growth. If a piece of a potato is separated from its parent plant it can survive the disjunction and—merely by continuing its growth—become a new individual which resembles the old. The reorganization is remarkable:—as remarkable as that of the fragments of growing crystals when they are separated and reconstitute themselves in the image of their parent.

It is, moreover, ludicrously untrue to assert that the

scientist cannot distinguish between a living bit of matter and the same matter dead. Most definitely is the living condition associated with certain electric and chemical potentials, with certain systems of surface-energies, with the presence of certain kinds of atoms in certain relationships. We do not know *all* the physico-chemical characteristics of living matter (if we did we could perhaps create it); neither do we know *all* the nature of any matter, living or not. As we chase the authors of the vitalistic fallacies from one corner to another—they are continually wriggling from the merely unknown into what they deem the unknowable—we may pause long enough to beat down another couple of ancient will-o'-the-wisps which mislead the traveller in these dangerous regions.

One is the "power of movement" possessed by living things. It is possessed also by non-living things, rivers, clouds, automobiles, occasionally mountains and continents. As for the "independence" of this power of movement-every student knows what that means to an ameba, how absolutely its motions are controlled by forces and substances around it. The other biological banshee is the statement that living things maintain themselves "in spite of the environment." This has a certain superficial plausibility.

Plants and animals do seem to "overcome" obstacles to their continued existence, to "solve" problems. In a pretentious and only occasionally unsound popular survey of the living world we read: "Existence in the form of distinct individuals which... reproduce by a sort of inherent necessity is a third distinction between living and non-living things.... Drops of oil or water grow and break up under suitable conditions, but not through any innate disposition to do so.

Living things display an impulse to reproduce themselves even in adverse circumstances." Shades of the vital force! The division of a drop of oil is a matter of only environmental effect, the "innate disposition" of the material is of no consequence; a lump of lead would do as well, forsooth! Whereas in living things there is an inner "impulse" which brings about division, in spite of all the world.

The biologist who wrote this nonsense well knows that

the reproduction of many kinds of plants can be absolutely controlled by slight and arbitrary changes in their nutrition or illumination or temperature; that no matter how "adverse" the circumstances to *other* life processes, reproduction occurs only under the conditions which favour it. So does a locomotive "overcome" a steep grade: in spite of gravity and because of its inherent constitution! But deny it oxygen, water....

Years ago Hobbes asked: "Why may we not say that all automata (engines that move themselves by springs and wheels as doth a watch) have an artificial life?" The same thought viewed from the other end is the mainspring of modern biology: why may we not say that a living body is a sort of natural machine? This is the answer of the biologist to the question, What is life? The differences between animate and inanimate are legion; but they are matters amenable to scientific observation, which discloses as yet no cleavage between the two realms of nature. In fact, among the recent discoveries of biology are substances—the filterable viruses—which seem to tremble on the verge of life, which the chemist hesitates to pronounce dead and the biologist is unwilling to declare living.

True it is, most evidently true, that physico-chemical research can deal with such activities as occur in living bodies only a few at a time; whereas in a living organism they occur all together with a complex wealth of coordination that confounds the imagination of some philosophers (who have apparently never seen a big automatic machine, such as a printing press, in operation). But there is no single activity accessible to scientific method which marks a distinct boundary.

If we postulate such a function, we are forced to retract by the very progress of science. Organic substances were so named because men believed that the vital spirit was necessary to their synthesis—now a commonplace of the chemical laboratory. Digestion was recently taught by eminent scientists as a peculiar manifestation of vitality; this chemical change is now easily demonstrated, in lifeless test tubes, to elementary classes. "The order of nature," said Huxley, "is ascertainable by our faculties to an extent which is practically unlimited."

A compendium of knowledge, backed by the prestige of professors of biology and education, informs us that "the most obvious thing about plants and animals is that they are organisms; that is, that their bodies are organized for a purpose." If this is true, then it constitutes an amendment to the conclusion just reached: living and dead substances are indeed alike in constitution and in energies, but differ in organization.

As one might say, a locomotive indeed contains the same materials as are found in the mountains over which it moves, but differently organized, compounded so that they serve a purpose. Certainly this seems plausible. So obvious does it seem that living things are "organized for a purpose" that perhaps the commonest question asked of a teacher of biology is: "What is it for?" Why does a tree have wood? Why are its cells elongated? Why does a root form hairs? Why is the skin of a leaf waxy and why does it have stomata? Why are the petals of a flower so bright and why do they exude nectar and sweet odors? What is a flower for? What answer can science give to such questions?

Cover a sprouting potato with a box into which light is admitted only through an opening in the side. The shoots curve in their growth, bend unerringly towards the opening, towards the source of light. If temperature and moisture permit, they grow through the opening and emerge into full illumination; there to display their leaves, evaporate water and manufacture food and oxygen.

You ask: "Why do the shoots bend towards the light?" I can answer: "Because they need light. The purpose of a plant is to grow to maturity and to leave offspring and this it cannot do unless it can obtain light. It is obvious that if the shoots had *not* bent as they grew, they would have perished miserably in the darkness."

If that does not content you, I can go on to discourse of the auxins, those substances which stimulate the enlargement of their cells. It is known by actual experiment that auxins accumulate on that side of the stem which receives the least light. The cells of that side therefore enlarge more rapidly than those on the other; the result of such unevenness in growth is naturally a curvature, a bending towards the side that is best lighted.

We have answered the question in two different ways, which are not wholly reconcilable. The known behaviour of auxins accounts for the curvature of a stem without invoking the needs or purposes of the plant. It can be extracted from a stem-tip and reintroduced on the best lighted side of the stem and then causes a curvature in the wrong direction. How do the purposes and the mechanical causes work in partnership? Or is one sufficient without the other? Do not try to evade the question by saying that the purpose of the plant produces the auxins; for we shall find mechanical causes then for this step also in the chain of causation.

The underpinnings of science, of all connected thinking, are statements of facts; *mere* facts, what has been called the "plain tale." Science has been defined, in words which have not lost their pungency in these many years, as *the knowledge of consequences and dependence of one fact upon another*. When we connect facts we call them "cause and effect," in the order in which they occur.

By such a process do we identify auxins as the cause of stembending; additional facts (not as yet completely known) may be in turn assigned as the causes of auxin manufacture and migrations and so on. On the other hand, let us not forget that the stem actually does *succeed* in reaching light and only by virtue of that success continues to live and prosper. If we keep in mind the well-being of the plant, we may describe it as the ultimate end or goal: what Aristotle called the *final cause*. Without such causes, indeed, is not life without meaning and biology a delusion?

So throughout our delving into nature we are guilty of a double meaning when we ask "Why"? We are satisfied sometimes with one answer, sometimes with the other; rarely do we insist on both. Why does a leaf manufacture food? Because food is necessary to the life of the plant; also because carbon dioxide and water encounter each other there in a setting of certain circumstances. Why does a stem form long tube-like vessels? Because it must carry water to the leaves and water flows readily through tubes; also because of certain innate and inherited features of its growth-mechanism which result in that kind of a structure.

We need not limit our examples to *living* things. When a ray of light passes at an angle from air into glass, it is bent and somewhat retarded in speed; the whole path which it takes from its source to a given point being of all crooked paths to that point the one which it can traverse most rapidly. Its behaviour is governed by the nature of light and the structure of matter; its purpose is to reach its destination in the shortest possible time. Rivers form valleys and, so doing, become straighter.

The causes of such changes are to be found in gravity and in the properties of the crust of the earth. The purpose of the rivers is to reach the sea as quickly as possible. The terrible storms known as tornadoes result from certain atmospheric conditions encountered in the summer on the great plains; partly from the imprisonment beneath a blanket of cool air of a layer of overheated air next the ground. The purpose of tornadoes is the chastisement of men for evildoing and following after false gods.

"But no one is seriously concerned with the purposes of tornadoes and rivers; purposes belong to living things." It is true that the purpose of a storm or of a stream smacks of the supernatural, if not of the absurd. A chicken runs across the road in order to reach the other side; but a stone follows it only because of the slope of the surface and without a purpose. However, the purposes of rivers are introduced apparently in good faith into at least one textbook of physiography. And, after all, is not the author within his rights? Is it not merely egotism which insists that living beings have purposes which inanimate nature lacks?

The founders of that Mediterranean civilization from which our own is so largely descended personified nature quite freely and without shame. To the Greeks of olden time every river, every crag was the abode of a spirit, with whose existence it was to some extent identified. The actions of natural bodies were explained by the wishes, bad temper, or carelessness of the deities concerned.

Pestilence was due to the malice or vengeance of a god and was effected by arrows shot from the divine bow. Winds were the breath of immortals, who were kept in a bag by a superior immortal. The trees were peopled by spirits whose aid must be invoked by shepherd and farmer. Since natural events were at the mercy of so great a throng of capricious beings, control over nature was possible only through a technique of propitiation: magical or religious rites stood to them in the place of science—with varying results.

As for us, a more or less monotheistic religion and an increasing technical facility have changed our attitude towards surrounding nature. We look for "natural explanations." We personify chiefly when we are moved:—a mechanic will apostrophize an engine that will not run and the eloquence of fear speaks in such phrases as "she's a bad one," uttered of an approaching storm.

We refer also to things we love, such as ships, by personal names and personal pronouns. But we learn to control mechanisms by understanding those sequences which we call causes and effects, not by trying to work on their feelings. We have even extended this attitude into our dealings with plants and animals and certain psychologists, fashionable a few years ago, taught us to exclude purposes even from our interpretation of *ourselves*. It is a far cry from the age of Pericles to that of Watson and surely all possible attitudes may be found between these two.

At least it should be clear that our ability to deal with this world into which we have stumbled, to shape it perhaps a trifle nearer to our hearts' desire, must rest more upon "knowledge of facts and the dependence of one fact upon another" than upon an approach to natural objects as persons actuated by purposes. It is not necessary, indeed it skills nothing, to account for the sprouting of a potato in terms of its foresight and cunning.

But this is not (as certain rather naive teachers of botany would have it) the end of the matter. Is it *not true* that things may be moved by purposes even though they are unconscious of them? Lest this be thought mere wordjuggling, let us hasten to a distinction. If your excursion this May morning is motivated by a conscious desire to gather specimens of flowering plants, there is no doubt of the existence of something in you which you may call a purpose; not only that, but it is your own purpose, *a private purpose.*

We hesitate to ascribe such a motive to the motions of a leaf or a river; it seems fantastic to credit them with desires and pleasures. But inanimate machines may certainly be said to have purposes, in a different sense. The purpose of a watch—is it not obviously to tell time? If such an expression is permitted, it can mean only that the machine was designed by a man or by society animated by that purpose; its purpose is imposed upon it from without, it is an *exterior purpose.*

A watch can act reasonably although it has no reason of its own. It is in such a sense that even a tornado may have a purpose and it is a rash biologist who would deny that any living thing, man or moss, serves purposes. "Nature makes nothing in vain," quoth the philosopher. Purpose may be inherent in life and living things unintelligible without it.

Some such path of meditation has led to a remarkable modern theory, which one of the most eminent of zoologists has called "the declaration of independence of biology." Its proponents cry that only by grasping and using the sense of this charter can biology progress to further understanding of life. If this is so, it is imperative that it be given careful thought by anyone interested not only in what biology has revealed in the past but what it can illumine in the present and in the future.

We are told that while living matter contains no atoms and is moved by no forces absent from dead matter, the relationships which exist in it are peculiar to it and, more than that, could not have been predicted from any possible knowledge of matter which is not alive.

The living qualities are said to *emerge* rather than to result from the antecedent conditions. So when two chemical elements are compounded, the new substance has new qualities which are not traceable to the characters of each substance taken separately. Two gases, hydrogen and oxygen, combined under appropriate conditions, give rise to a liquid, water. The new quality, liquidity, *emerges* from among the qualities that existed previously but is not predictable from them.

If we were thoroughly acquainted with the nature of hydrogen and oxygen in the gaseous state but in no other, we could not foresee, on the basis of all our knowledge, that an entirely new character would appear when they were joined. Perhaps this crude example is fallacious; but it will serve to represent the thought of the theory of emergence. At a certain point in the history of the world life is said to have emerged as a new kind of activity, a new level of phenomena, not wholly describable in the terms used for non-living matter. The organization or power of organization of organisms and the purposes which they are said to manifest are qualities peculiar to the living condition, not traceable to the known or knowable laws which describe the behaviour of nonliving atoms.

Here evidently is a new kind of vitalism. Life is distinguished not by a special force or a special matter, but by purpose; we return to the statement previously quoted of "the most obvious thing about plants and animals."

Let it be admitted first that this theory cannot be thoroughly treated here. It would put too great a strain upon the patience of all but the budding philosopher. It is a theory of knowledge and treads on the edge of some of the most difficult country of all philosophy. He who is curious to explore may first ask: Is it indeed true that if you knew *all* about two gases you could not predict the result of their union? It is quite possible that at least you might be able to predict *all that science can reveal of the new character*. Of course you cannot predict a sensation which you have never experienced; but this leads us only into a discussion of what you mean by prediction and whether it is necessary to causation.

Let us rather, unskilled in swimming, not venture outside

the ropes which limit the shallow waters. Let us grant, with magnanimity, all that emergence theory claims and agree that living things are instinct with purpose. Let us now ask how this contributes to our real understanding of living things, how our science of biology is augmented. What new vision of truth is revealed by the lifting of such a curtain?

Purposes in the abstract obviously do not help us in actual work with living things; we must know what purposes. It is here we find trouble, for how may we recognize them? "The purpose of a living thing to keep going," says a philosopher, "is plainly a fact." Perhaps so; but does that help us to understand why it dies? In our ordinary intercourse we infer the purposes of others from knowing our own; or, if we are in doubt, we ask and may be answered. Since it seems profitless to inquire of the Author what are the purposes of living things, since it is surely unsafe to infer the purposes of a tiger or of a toadstool from our own;—is there reason for believing that we can recognize them from the behaviour of such organisms?

The fact is that when we attempt to do so we are merely transferring desires of which we are conscious in ourselves. We feel attached to the continuance of life and we act so as to further that end. Since other creatures also live, since some of them evidence something comparable to our own desire to live; since their life frequently depends on a very complicated mechanism which is very difficult to understand in purely descriptive terms; we say "plainly they also are kept going by their purpose to live."

Whether this is true or not is, strangely enough, beside the point;—which is that such pure subjectivism, egotism, has no place in science. *We cannot actually recognize purposes from actions*. If because an organism lives we are to infer a purpose to live, then when it dies (and it invariably does so) we must use the same logical procedure and speak of its purpose to die.

Is this absurd? Perhaps the view expressed so often in the Old Testament is at least as plausible as any suggested by modern philosopher: the Great Designer may be continually thwarted in His purposes by the actual behaviour of His creatures. Or, to adduce another system of belief, may it not be the inherent purpose of living things to die as soon as possible, a purpose in which they are temporarily hindered by circumstances which keep them alive? If they could but subdue their appetites Nirvana might soon be realized.

We, poor worms, think that everything must have been purposely designed to do what it does, because, forsooth, we have designed a few things ourselves—which often do things which we did not anticipate. Was it the purpose of the inventors of automobiles to kill and injure thousands of persons, or to ruin those whose income is derived from railways? Strange it is also that those who labour to introduce purpose into biology do not realise that if they succeed they must reintroduce it also into physics, chemistry, astronomy and geology. A mountain is obviously designed to bear waterfalls and a cloud to irrigate the soil. Could anything be better planned than the moon for causing tides or liberating tender sentiments in human hearts (whichever purpose may have been intended)?

It may be admitted that, as some writers are fond of insisting, to suppose that a plant is animated by the purpose to survive may sometimes lead to an understanding of its structural peculiarities. One can assume that a plant which grows normally under water, having no need of internal stiffening to hold it erect, may lack the fibrous cells found in stems which grow on land and that, since it is likely to suffer from a limitation of the oxygen supply, it should have internal chambers which store gases and act as channels for their diffusion.

Both these suppositions are true of many aquatic plants. But such a method of attack leads to false conclusions almost as often as to true ones. "We should keep in mind," wrote Charles Darwin, "the obvious fact that the production of seed is the chief end of the act of fertilization." In such a vein he continued to his dogma that "nature abhors self-fertilization."

The generalization is obviously medieval. Nature was once considered also to abhor a vacuum; but the explanation of the rise of a liquid in a closed tube from which air has been exhausted actually does not refer to the likes and dislikes of the Great Mother. A contemporary of Darwin pointed out that (presumably in spite of nature's disapproval) a number of very successful species are propagated exclusively by self-fertilization. It is also said in many textbooks that cockleburs are provided with hooks "so that they may cling to the pelts of passing animals and so disseminate the species." The fact is that the burs thus translated frequently land on hilltops where they cannot form a new colony; while the successful dispersal of the species (as every midwestern farmer knows) is by water—the vicious and apparently purposive hooks serving for nothing useful.

Some years since I sat in an "educational" programme in a London theatre and watched a moving picture exhibit the marvels of plant life. One of the captions of this production, which was under the auspices of a governmental department of education, was as follows: "In order that the plant may have water, the root hairs must absorb more water than is needed and the leaves therefore have to give off the excess. This probably also helps draw up the nourishment." The plant seems to be working at cross purposes! And nothing could better demonstrate the unreliability of explanations beginning with "in order that." The leaves of a plant cannot help giving off water most of the time, excess or otherwise and experiment has shown that the water does not "draw up" the nourishment.

More than a century ago Thomas Nuttall, intrepid and eccentric botanist and explorer of our western plains and mountains, wrote as follows: "The tubes of this species [a pitcher plant]... are commonly crowded with dead flies and other insects, perishing in imprisonment by one of the wonderful but simple accidents of nature;—a lesson for the incautious!—but no proof of instinct or necessity in the passive Sarracenia which could probably well maintain its vegetation without the aid of dead insects, a remark equally applicable to many other plants which accidentally prove fatal to insects, such as the wonderful Dionaea, which in its native swamps as frequently catches straws as flies and will equally enfold anything, so subject is it in this respect to the blindness of accident. Still in the ascidia of the Sarracenia there appears to exist no ordinary degree of ingenuity to accomplish a purpose apparently of small importance to the plant itself...." Nuttall's science was as good as his English was uncouth; but what a tremendous effort was necessary, what clumsy mental squirmings, to pierce the prevailing fog of superstitious regard for the divinity and cleverness of all living things!

For an antidote to all the amiable but muddle-headed romanticism so characteristic of our age read an essay by that penetrating naturalist Richard Jefferies. A bird builds its nest in blackthorn or hawthorn which yields it food and a convenient wilderness of branches and which to it offers no obstacles, no difficulty of access; it does not anticipate that a human nest-robber will have trouble penetrating the bristling screen of thorns. "To understand birds you must try and see things as they see them, not as you see them."

A well known zoologist, writing for the general public, speaks of birds "being sensitive to red and whether they actually prefer it, or whether the flowers that set out to attract birds have developed a colour invisible to insects so as to remain unvisited by creatures to which they are not adapted..."—one could wish that this countryman of Jefferies would read him.

The rose-coloured spectacles of teleology led another scientist to write that "plants are only sensitive to what will help them in their growth and development." They are immune to death? In the most sumptuous popular biology of recent years the confused thinking is betrayed by such sentences as this: "Of the grubs that hatch out she gives one much more food than the rest, in order that she may have a worker to assist her as soon as possible." O clever ant! It is not true that feeding is the differential factor in the production of the different types of ants in a nest. The authors of this glorified textbook dismiss the idea of a *divine* purpose with the extraordinary argument that evolution has produced much that is bad! Thus does theology itself turn out to be a branch of biology; or vice versa?

Misled by the writings of biologists, even philosophers

fall into error. The familiar response of a plant to gravity is often adduced to show that a mechanical explanation of life-phenomena is impossible; for the root responds to gravity by growing down, while the stem of the same plant responds to the same stimulus in an opposite manner. Hence (they argue) there is some other factor, imperceptible to our means of observation, which determines the behaviour of the plant.

The difficulty originates in the tacit assumption that the plant is a unit and uniform. If you begin by believing that it has a purpose, of course you are obsessed by the notion of a plant as an autonomous body. If you first study the facts, it is clear that a plant is a sort of republic of parts, normally harmonious and correlated, but differently constituted. There is nothing mechanically difficult about the difference in the responses of root and stem:—have the gentlemen forgotten that the two ends of a piano respond differently to the same stimulus?

One of the most naive arguments for the introduction of purposes into biology is that the language of this science is teleological. Vessels carry water to the leaves *for* photosynthesis; plants absorb from the soil certain inorganic substances which they *need;* most plants *require* free oxygen if they are to continue to live. The criticism is just; but it is only a criticism of the language. It is true that a wholly consistent botanist should speak of the life of a plant only as a result of the presence of certain substances; should avoid saying that plants require oxygen (whose concern is it whether they live or die?) and instead make it clear that they live *because of* the oxygen which penetrates their tissues.

This is actually the point of view which animates botanical research. It is a matter for some regret that botanists are so careless about language that they simply do not trouble to make their thoughts plain—even to themselves. Hence the misunderstanding of their methods by those who "teach science" in the schools; hence also the muddled philosophies uttered by very eminent biologists in their declining years.

The traveller along this tortuous and often dimly marked path has had many a glimpse into the morasses and quicksands which we have skirted. Have we now circumvented these dangers only to emerge into an and land, a world without purpose, without soul, without beauty, without hope? It is indeed this dread that has detained many of our fellow wanderers. Any sensitive mind which has been subjected to an overdose of "behaviourism" is apt to recoil into the ample if somewhat amorphous bosom of teleology.

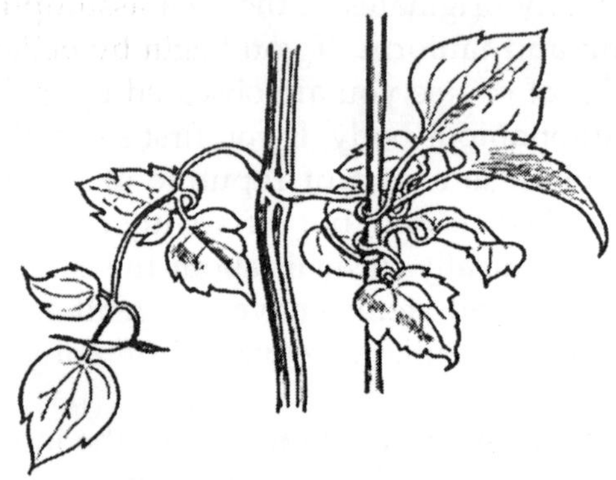

Fig. Tendrils

It is not my purpose to demonstrate that there are no purposes in living things. I am concerned here only with the description of nature and—to limit myself further-with that species of description called scientific; with nature seen by the methods of the laboratory. None but the fool denies his own soul. But the beauty of'living things, their personalities, their desires and ends, are not among those findings called scientific. The chief justification of such a limitation is:—it works. The methods of the laboratory those methods always called, for some reason which I do not entirely grasp, the methods of physics and chemistry—have yielded great insight into life-processes and a remarkable improvement in our control over life; results which the biologist, with a strange bashfulness, attempts to conceal in a cloak of medieval folk-lore and superstition. They have given us knowledge for which our forefathers would have hanged or burned us, in the conviction that it could proceed only from an alliance with the Adversary.

Trust not the biologist who pretends that the actions and

structures of living things are "explained" by their purposes; or who talks of introducing purposes into the data of his research. Trust him not until he shows you a purpose, *properly isolated and identified,* in his laboratory. If he cannot do that his assertions have as much value for science as the conviction of the Duchess that "Everything's got a moral, if only you can find it."

As for the declaration of independence, ask a physical chemist if he would like physics to be independent of chemistry and his own science of both. Facts are facts, obtained in any way that human ingenuity can devise. Whether they relate to gases or to mountains or to jellyfish does not affect their status as facts. The methods of investigation are a true commonwealth, shared by all competent to use them; regions can be labeled "chemical" or "geological" or "biological" only in recognition of the interests of those who work in them. Does not "independence" really mean a turning back to the language and methods of thought of the dim dawnings of biology; to the tales of its scientific childhood, with Emergence as a new fairy godmother?

Chapter 6

Flowers and Their Fruits

Four or five thousand years ago when an Egyptian gentleman called upon a friend he was greeted by a servant who offered him a lotus flower. This he held gracefully before his face as he conversed and its sweetness embellished his discourse; when it wilted, another was provided. At banquets in imperial Rome the diners wreathed themselves with roses and decked the walls with garlands (to which our pallid plaster decorations still testify); rose petals drifted softly and sweetly through the air. Medieval monks tempered their austerity with beds of roses and lilies in the cloisters; flowers adorned the great church on feast days, to the glory of God. Roses, dyed as if with the blood of martyrs, were symbols of the Divine Passion; lilies of the purity of the Virgin Mother.

Humility and constancy, the sorrows of this world and the joys of the next, all were expressed by flowers. In India the rose gave birth to the goddess of love and beauty; in Greece the beloved of a god, dying, was transformed into a hyacinth. Lotus flowers gave form to capitals of Egyptian columns, roses to patterns woven in Sweden and knotted in Bokhara.

"Flora, with a full lap, scattering knowledge and flowers together; everything good and sweet seems to come out of flowers, up to the very highest thoughts of the soul and we carry them daily to the very threshold of the other world." We open our eyes upon a world bedecked with flowers and flowers surround us when those eyes can see no more.

The variety of flowers is both pleasing and overwhelming. If we call to mind the images of a few common ones—roses, lilies, daisies, tulips, sweet peas, daffodils, dahlias—we cannot fail to be struck by the diversity in their form, in their various colours, their sizes, their odors. What is common to all of them that we name them all flowers? Only bright colours and sweet smells? Certainly some flowers almost justify the cynic's definition: "Flowers are commonly badly designed, inartistic in colour and ill-smelling."

A flower of Aristolochia, the Dutchman's pipe, smells like carrion; skunk cabbage needs no further introduction and the unbeautiful flowers of ragweed (scarcely recognized as such) excite in us only a feeling of disgust. But all these are truly flowers, as truly as roses and lilies are flowers. Flowers are to be known by something other than the qualities for which we prize them in our gardens and homes. To this miscellany of beautiful and repulsive and conspicuous and insignificant objects there is a structural pattern and a functional unity which have little to do with colour or odor and are independent of human sensations, symbols and ideas.

Fig. Tulipa

Many flowers are surprisingly small. For first acquaintance with the general theme, it is requisite to choose a large one. A tulip is convenient; its parts are well known and easily seen and sufficiently typical. There are first six large coloured overlapped parts which form the bright and graceful cup. Such coloured parts are generally known as petals; the word is unfortunately used in a somewhat restricted sense by the botanist and so is apt to lead into confusion. More technically the outer floral parts are the perianth. The perianth of a tulip is arranged in two rings or cycles of three parts each, the outer three overlapping the inner and covering them in bud.

Within the perianth other parts may be seen: dustyheaded stamens, six of them and the sticky and convoluted head of the pistil. The Latin word for meal, flour, is pollen; pollen is formed and discharged by the stamen heads. The pistil is named for the apothecary's pestle.

All the members of the flower are attached side by side and close together to the slightly swollen tip of the flower-stalk, best seen in a dissected flower. From this torus the sap- and food-carrying bundles extend into perianth, stamens and pistil and nourish them. The perianth grows from the margin of the knob of the torus, the stamens are attached next within the perianth and the pistil occupies the centre.

The pistil is central in function as in position; it is the most characteristic part of the flower, without which indeed flowering would have no meaning for the species; for within it is generated the progeny of that plant. It has a somewhat distended and more or less three-angled prismatic lower part named the ovary, which is tipped with the three-parted viscous stigma. If the ovary is cut across, three cavities are disclosed, running side by side and lengthwise in it; each contains two columns of minute greenish bodies. Here is the desired result of our prying into the inner recesses of the flower: these delicate small objects, named little eggs or ovules by the botanist, are the future seeds from which a new generation of tulips may grow.

For it is true, as John Ray long ago pointed out, that, while all things exist for the greater glory of God, plants do not live for the sake of man. Their lives are passed and their work is done in disregard for the wants of any other living creatures. To gain an understanding of flowers which surpasses the merely horticultural, one must inquire yet further into the minutiae of stamens and pistils.

These are called the "essential" parts, to whose activities the lovely perianth is more or less accessory and sometimes unnecessary. The many flowers in which stamens and pistils fail to develop must be regarded as horticultural freaks. What though the dahlia and the peony are the most magnificent things in the garden? Left to themselves and denied our watchful care, our assistance in propagation, they would quickly perish from the earth, or survive with much less of ostentation; the redundant blooms which we prize are but a useless encumbrance to the living organisms.

At the other extreme are things which the botanist calls flowers and which are scarcely visible to the observer; a flower of Euphorbia may consist of nothing but one stamen on its torus. Yet does this lowly flower add its fertilizing dust to the river of life; its existence has value to the species to which it belongs, little as it means to the gratification of human senses.

Fig. Ranunculus and Buttercup

A curious feature of flowers, already discernible in the tulip so baldly described above, is their numerical precision. A tulip has six perianth-parts, in two cycles of *three* each; it has *six* stamens, similarly disposed; it has a pistil formed of *three* parts united side by side so as to make an ovary with *three* cavities and a stigma of *three* branches. Tulips with other numbers of parts are seen, but infrequently; rarely enough so that they are regarded as monstrous. Arithmetical patterns are characteristic of most of the families of flowers, as old

Linnaeus, father of botany, long ago pointed out. Some flowers, indeed, are more vague in their requirements of parts. A buttercup centers on a cluster of small simple pistils, which grow crowded on the tip of the torus; their number varies even from flower to flower on one plant. Around them is a crowd of stamens which we trouble not to count. Roses, strawberry-flowers are similarly prolix; cherry blossoms have but one pistil but many and inconstant stamens.

But a geranium has ten perianth-parts in two rings of five; ten stamens in two rings of five and five pistil-segments united to form a compound pistil which is crowned by five stigma-branches uplifted on a slender style. In a dogwood flower we number the parts by twos and fours, in that of a mint by fives, fours and twos; a knowledge of these numberpatterns is an elementary tool in the classification and identification of flowering plants.

In most flowers also, as in those just named, the perianth is more differentiated than that of a tulip. The five green outer perianth-leaves of a geranium differ sharply from the five coloured inner parts: differ in size, shape and texture, as well as in colour. The outer cycle has been known since the time of Pliny as the calyx, the "coverer" (since in the bud it covers the inner parts). Its separate members are the sepals. The more decorative inner cycle is the corolla, the little crown; composed of the petals.

It is to be noted that the distinction between these whorls is not one of colour and use; in fact the disposition is sometimes the reverse of that just explained, so far as colour goes. In a larkspur the brilliant blue parts for which we give the plant place in our borders, these are the outer ones, the sepals; the petals are within them, smaller, often blackish, hairy, forming the "eye" of the flower and given mostly to the secretion of nectar.

Botanists have come gradually to a realization that petals and sepals, corolla and calyx, are indeed different in structure and perhaps in origin, as well as in position.

It is possible that Goethe, who by his persuasive eloquence confused botanical thought for a century,—it is possible that he was right in supposing sepals to be transformed leaves. Certainly some plants, such as peonies, have every gradation between characteristic leaves and the sepals; it is hard to avoid the conclusion that a sepal is a leaf-base, much expanded, which never develops the terminal parts. Petals, on the other hand, at least in some flowers, seem to be related to stamens, which are probably greatly condensed branches; certainly there is every gradation in a waterlily between stamens and petals and in many other flowers also.

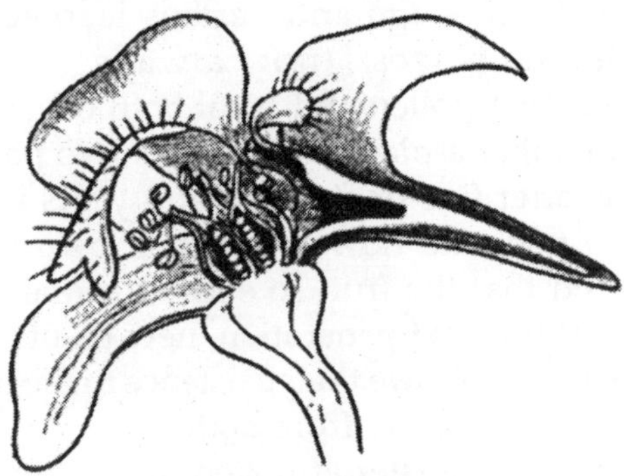

Fig. Delphinium

When we turn to the operation of the flower, to the working of its mechanism, we draw upon a heritage of knowledge which has accumulated over thousands of years. The first horticulturists, who lived in the valley of the Nile, learned that to have dates it was necessary to hang flowers which produce pollen among the branches of trees which do not; for flowers of date-palms are formed of either stamens or pistils, not both and the two kinds grow on separate trees.

The pollination became a ritual in the country of the Tigris and Euphrates and we have in our museums the majestic sculptured angels of that region, shown dusting with pollen the fructiferous branches. Without benefit of microscopes the pollinators were ignorant of the mechanism whose operation they were assisting and became fanciful in their attempts at explanation. Indeed among the Greeks and among the medieval scholars who copied them so dully, "male" and "female" qualities were assigned to plants without reference to their actual reproduction or to the stamens and pistils which may in truth be likened to the male and female parts of animals; an entire species was called "male," another "female."

Linnaeus and his contemporaries perceived, somewhat vaguely, that pollen and pistil must unite for the production of fruit. So little understood was the process of reproduction that it was long thought that the pollen developed into the seed, the pistil being a sort of nutritive receptacle for its development. This was still being taught in universities less than a hundred years ago and various learned societies in Europe offered prizes (not always awarded) for demonstrations that pollen and pistil both contributed to the next generation, that a plant actually has two parents.

Belatedly, after five or six thousand years in which men did not fail to pollinate flowers for their own purposes, we have recognized that the fruit so elicited contains (normally) seed in which is the next generation, new plants in miniature and that these embryos owe their existence to the union of their parents and inherit equally from both.

Pollination, the transfer of pollen from stamen to stigma, is accomplished by a variety of means other than the hand of man. In some kinds of plants merely gravity or contiguity suffices, as the wind swings the flower on a slender stalk. Some pollen is shed into the air in such quantities that it adheres to everything for yards or miles around. If such dust reaches our respiratory passages and there causes the inflammation popularly known as "hay fever," it is not remarkable that some of it also reaches the pistils of that species which produces it, whether in the same flowers or in others. The stigmas of such flowers are generally large, often branched and feathery; well fitted to comb drifting pollen from the air.

But of all the wonders of a flower perhaps the most wonderful is its cooperation with another organism; not even another plant but an insect or occasionally a bird or even mouse or snail. The sweetness of flowers has after all a biological meaning; animals lured thither by it and finding there sweet nectar as a reward, are the involuntary pollinating agents and the flowers, lacking wings and legs of their own, make successful use of those of other creatures. It is worthy of note that the wind-pollinated flowers usually lack bright petals and sweet odors.

The details of the apparently ill assorted partnership between flowers and animals are so remarkable that they have fascinated biologists and the unscientific public for many years. The kinds of flying insects are legion and the "contrivances" of flowers which make use of them are also numerous. The first requisite is a signal or flag which will be registered in a winged creature's sight, to which he will direct his course: coloured petals (or other flower-parts) bring dayflyers, white ones the moths and beetles of dusk.

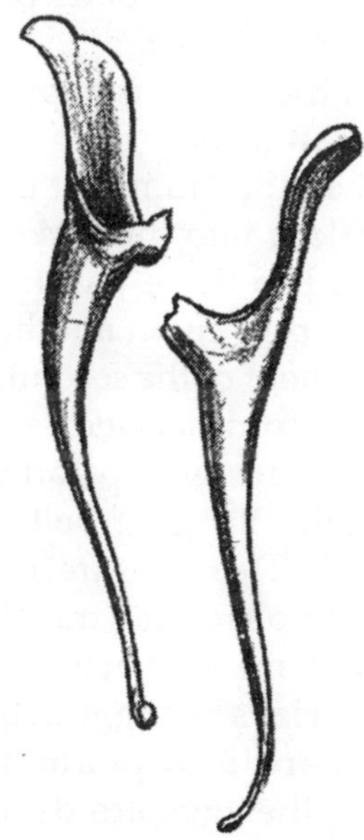

Fig. Columbine Petals

Then there is nectar set out for the visitors; so placed—often in pouches or spurs—that to reach it they must stumble or creep through or over the stamens and brush against the stigma, smearing themselves with the sticky pollen and wiping some of it off in the proper place. The stamens often are very numerous; in such flowers (an apple blossom, for example) different cycles of stamens may mature at different times, one set springing up as another withers down, so that a continuous supply of fresh pollen is on hand for several days. Some flowers hold the stamens under such tension that a slight touch releases them with a jerk; the motion sends out a cloud of pollen which dusts the disturbing body. This is a trick practised by the butterfly flower (Schizanthus) common in our glasshouses and by mountain laurel (Kalmia).

If we study the insects instead of the flowers we are equally impressed by the apparent care with which their parts are arranged so as best to pick up and transport pollen. On the leg of a worker bee is a small pouch in which pollen is accumulated. On another leg is a spur so placed that with it the bee can scrape the pollen out of its pockets and deposit it in its appointed place in the hive.

Of course much of the pollen is appropriated by the insect for the uses of the hive. But some is bound to be left here and there on a stigma; not all the grain that comes to the mill goes into flour, nor all the flour into the sacks.

We are accustomed to those natural dispositions which enable one organism to prey upon another, to the satisfaction of the first and the detriment of the second. The noteworthiness of the arrangements here described is that they serve two interests; they suggest partnership rather than pillage. The flower of Yucca, a white hanging bell of the lily family, is visited by a certain moth. This creature, crawling over the long stamen-head, takes up pollen and transfers it to the stigma; pressing it firmly into place as if actuated by tender concern for the plant world; then lays her eggs in the ovary of the pistil. While, thanks to this artificial pollination, the ovules are developing into seeds, the eggs are developing into worms, which eat some of the seeds.

The act of pollination is now seen to be not so disinterested as at first appears; it provides for the food of the larvae. But how did the moth know that? What guided her in the selection of pollen, in the transfer of it to precisely the right place for its germination—not, be it observed, the same place as she is to deposit eggs. Be it observed also that the plant is not extinguished by the attentions of the moth. There are not enough worms to eat all the numerous seeds of the Yucca pod; some escape, become new Yucca plants, which will nourish yet more moths.

A further peculiarity of many flowers seems contrived to get as much work as possible out of the involuntary pollinators. In hundreds, thousands of species, stamens and stigmas, though both present in each flower, do not mature at once in it. The pollen cannot germinate on the stigma of the flower which forms it, for the stigma is not "receptive" while it is being shed. The stigma cannot be pollinated by the stamens of that flower, because their pollen has already been shed or is not yet ripe.

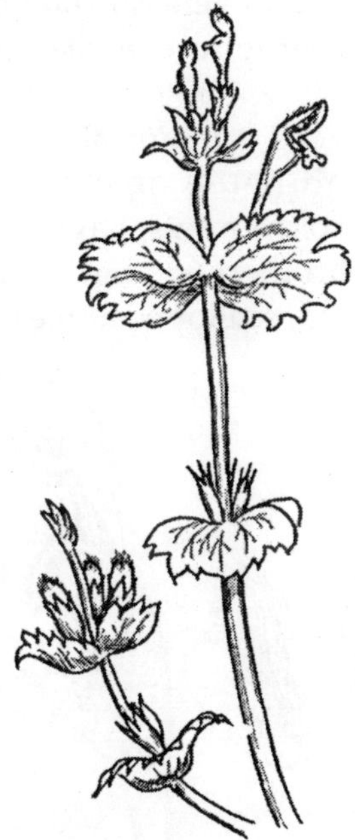

Fig. Lamium

It is curious and suggestive also that plants which indulge in wind-pollination so frequently bear their pollen and stigmas in separate flowers and even on separate plants. There are indeed so many circumstances which seem to compel or at least to favour cross-pollination, to render self-pollination impossible or difficult, that many botanists have been led to believe that this is one of the "obvious purposes" of nature; or that crosspollination is inherently superior to self-pollination. Darwin became dogmatic on the subject and many have followed him without sufficient criticism.

The first conclusion is without meaning for us as students of science; the second is partly true, but only partly. There are species continually self-pollinated generation after generation, without apparent ill effect. But it is certainly true that the structure of most flowers favors crossing. In many species this result is accomplished by mechanism of an ingenuity and precision that makes a purposive bias to our language almost inevitable and a teleological tinge to our thought pardonable. Flowers are so diverse that general statements are too vague; a few examples will serve to exercise the imagination and credulity of the reader.

There is a small weed common about paths and buildings, called variously dead-nettle or henbit; more accurately, Lamium. It is related to the mints, having the square stem and paired leaves of that brotherhood but not the odor. The leaves are dark green, rounded, scalloped at the edges, lacking stalks.

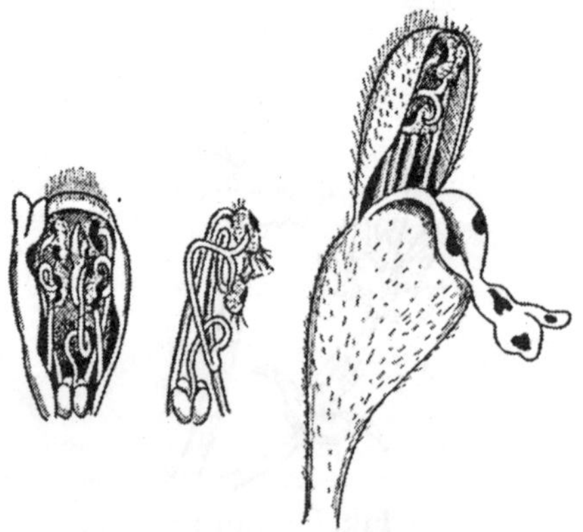

Fig. Lamium Petals

In earliest spring appear clusters of minute buds at the bases of the leaves, each a fivetoothed green cup in which may be seen a tuft of vivid magenta hairs. The cup is the calyx, the coloured hairs crown the closed corolla. This flower disappoints, fails in its promise of adorning the spring; it never opens. It remains merely a minute spot of brilliant colour almost hidden among the dingy leaves and finally withers away.

Yet later, when the corolla has departed, a fruit may be found in the calyxcup; four small nutlets, each containing a good seed. In size and shape this flower scarcely compels our attention; in the merry-go-round of spring it is a light that fails. But, as is so often true of small things, study reveals a situation we had not suspected: a curious and interesting state of affairs, which solves the problem just suggested, how a plant can form fruit and seed without opening its flowers.

To see its parts it is needful to detach a flower and slit its petals (which are permanently joined in a tube) up one side; their ends are free and the corolla thus opened may be unrolled and spread out in the manner illustrated herewith. The stamens are now visible, in two pairs: one from the lower, one from the upper side of the flower. Their stalks are curiously contorted. The upper two carry their dark brown heads to a meeting-place just under the closed tip of the corolla, in the midst of the flower.

The lower stamens twist downwards and thrust their heads together just below the others. All these brown heads wear rather towzled white hairs. The ovary at the base of the flower is of four distinct lobes, from the midst of which arises the style, forked like a viper's tongue; the two prongs spread apart just between the upper and lower pairs of stamen-heads, one pointed tip reaching each pair. Such is the disposition of stamens and pistil inside the closed flower.

The style continues to grow longer, as if it would push its forked stigma out into the air through the end of the corolla; in vain— the corolla remains firmly closed, the stigma is held by the imprisoned stamen heads and the over-ambitious style must coil its excess length within the petals.

There are no insects here questing after honey and carrying pollen from stamen to stigma; no visitors are admitted. The stamens split open and exude a mass of orange pollen. Some of this comes in contact with the points of the style and promptly behaves as pollen normally does when it reaches a stigma; it sends forth microscopic tubes which bore their way down through the style into the ovary. Then that pollen which does not actually touch the stigma behaves likewise; its tubes find their way through the cavities of the stamen, through the mass of pollen, into the stigma. These pollen tubes, as in all flowers, are the means of fertilization; through them the pollen contributes its share to the coming generation. In Lamium also they anchor the stamens to the stigma.

Here is a flower which is independent of outside aid, of the winds of heaven or its winged creatures; which somewhat justifies Linnaeus' belief that the structure of flowers, which usually contain both stamens and pistil, results easily in selffertilization, in contrast with the usual history of animal reproduction.

But this is not all. After a few weeks, as the air becomes warmer and the days longer, you will see other flowers formed on the same plant, larger flowers which open their corollas. The petals are bright mauve, with magenta spots and hairs; they are partly joined to form a flaring tube with a hooded and two-lipped end. The hood is crowned with the tuft of gaudy hair; the lip is made of two small lobes; between hood and lip are two other narrow lobes.

The stamens stand up under the hood, in two pairs as before; as the flower opens the lower pair of stamens straightens up and carries its pair of heads to the top of the hood, above the other pair. This leaves one prong of the style projecting far from any pollen —facing away from all the pollen. The stamens display their sticky wares and some of the grains of pollen may find themselves in contact with the upper lip of the stigma. The rest of the story is a matter of chance and the casual visits of insects, who come seeking nectar that gathers in the corolla-tube.

To reach the sweet secretion, they must push against the pollen-masses and will carry some pollen away with them, sticking to their backs; this will easily be rubbed off against the stigma of the next flower they visit.

There is a precision, an apparently calculated ingenuity

here that recalls the operation of man-made machines. It is difficult (to say the least) to regard pollination as the accidental result of the structure of a flower and the independently formed habits of insects. All contingencies seem provided for. First, a supply of seed is ensured, independently of what is going on in the world.

Then, in the warmer weather, when insects abound, they are pressed into service for pollination. The whole scheme is dependent upon a multitude of small circumstances: upon the nice fitting together of the parts of the flower, the precise coiling of the stalks of the stamens and the pairing of their heads, the orientation of the two prongs of the style, the timing of the opening of the flower.

Lamium is by no means an extraordinary sample of the ways of flowers. To observe flowers is to see nature at work in one of her most versatile moods. The carrying of living dust from its source to its fruition is brought about in an infinity of ways, some almost unbelievable to those who think that only man can create mechanism. The explanation of the origin of such arrangements so far eludes our mechanistic explanations and leads us into personalizing both plants and animals—or to frequent invocation of the plans of the Creator.

Everyone knows the common violet whose five flaring petals wear their loveliness so modestly. Look closely at the solid green or orange knob in the centre. It is the end of the style; the stigma is in a little pocket on its lower side. Just beneath this the lowest petal is somewhat hollowed and the six-legged visitor can insert his proboscis into the spur of this petal where the honey awaits it. If there is pollen on the proboscis (from a previously visited flower), it is easily rubbed off into the stigmatic pocket.

The stamens are clustered in a cylinder around the pistil, their margins in contact; they open inwards, the pollen is consequently held between them and the wall of the ovary. The cylinder is even closed at the end by small orange scales which tip the stamens and form a cone beneath the knob of the style. Though separated by only a millimeter of ovary-wall from the ovules which await their fertilizing tubes, the pollen cannot germinate in its cylindrical prison. From the lower two stamens a pair of horns projects into the spur of the lower petal; these exude nectar, which drips into the spur.

Any intruding beak may strike these horns; this suffices to jar the stamens apart and release a shower of golden pollen which falls upon the disturbing body. As it is withdrawn, the beak encounters and closes, a tiny flap which hangs just behind the stigma; no pollen from that flower reaches the receptive cavity.

Fig. Viola

Fig. Viola Petals

The pollen so ingeniously contributed goes on to the next violet, there to be rubbed off in the stigma as the beak is thrust in. The needle deposits pollen on its way in, acquires a fresh supply on its way out. Here is a flower in which self-fertilization seems impossible. How then if insects are lacking, prevented by rain from flying? Actually the flower by the mossy rock often fails to set fruit. But, like Lamium, Viola has two strings to her bow. Other flowers are formed later on the same plant. They never open; they are scarcely noticeable beneath the heart-shaped leaves; in some species, indeed, they grow underground. The sepals have the usual appearance, but never spread apart.

There are no petals, or rudiments only. The stamens and pistil are much as in the betterknown open flowers; but the tip of the pistil is coiled completely around and backwards *within* the cone of scales which tips the stamens and bears the stigma at its extremity. The pollen sends its tubes through the walls of the stamen into the stigma, whence they follow its crooked path to the ovary; fertilization follows. Lift up the leaves of a violet in summer—after the blue flowers have long since passed away—and there stand the seed-bearing capsules on slender stalks.

What a world is involved even in a small flower chosen as an emblem of humility! We remark not only the structure of the flower, complex though it is; it is the fitness of the parts, their coordination in an apparently well planned conspiracy, that excites our admiration. Observe that small flap behind the stigma, which prevents pollen being scraped into the cavity from the inside of the same flower; this only in the open flower. Remember that coiled tip of the stigma —only in the closed flower. Consider the method of opening of the stamens, which is useless except for the projecting nectaries which must be in the way of entering probosces.

What if the upper stamens had formed the nectaries, or the closed flower developed a stigma which projected from the staminal cylinder? There are many other such flowers: many plants which luxuriate in two kinds of flowers, one seemingly especially designed for self-fertilization, the other for cross-fertilization. Darwin wrote a book about them.

But of all the prodigality of elaborate contraptions which bring about the cross-fertilization of flowers, the most remarkable are those displayed by the royal family, the orchids. These gorgeous and ornate creatures flaunt their often weird beauties most abundantly in tropic forests. Brought from their native land, they flourish in our greenhouses (if given expert and expensive care), multiply and hybridize and serve as an adornment of wealth.

Many small orchids also grow wild in temperate countries and various climates. The ladies' slippers or moccasin flowers of northern bogs rival in showy beauty their relatives of the jungles and on many a wild pasture or cool mountainside you may find the delicately fringed Habenarias. Even the tiny flower of ladies' tresses (Spiranthes) has the same complex type of structure that distinguishes a magnificent Cattleya or Laelia. The varieties of this structure are as numerous as the different tints of orchids and correspondingly the mechanisms by which pollination is effected are greatly diverse.

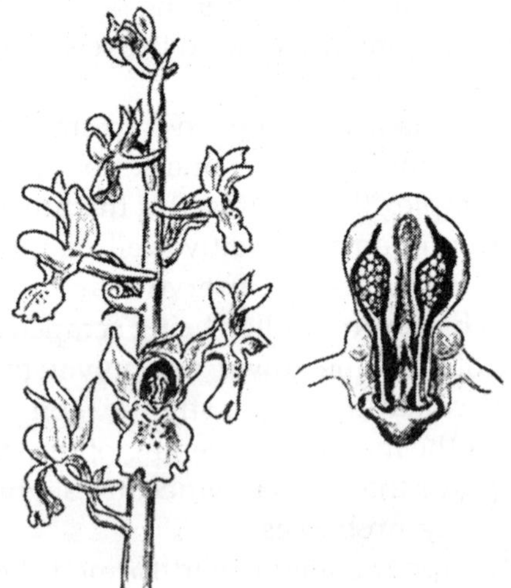

Fig. Orchis Petals

One of the first to be thoroughly described was the common wild orchid of England, Orchis mascula, studied by the great Charles Darwin. And since, though not large and splendid, it is almost as remarkable in its pollination as any of its more showy relatives, we cannot do better than follow the master's description.

The flower of an orchid is distinguished by an enlarged petal called the labellum or lip, which projects usually forward on the lower side of the blossom and takes a great variety of forms and colours.

It may be a large flat petal, or it may be fringed, cleft and crested. It may extend backward into a hollow spur. It may be distended into a pouch or slipper. It usually secretes nectar, which may collect in the hollow part; or it may itself be succulent and edible. The sepals and the other two petals are not usually remarkable save in their exquisite variety of colours. But just above the labellum stands the most characteristic and curious part of an orchid flower; named the column.

It is made of a stamen (two in some orchids) united with the style of the pistil, the ovary of which is below, hidden in the base of the flower. The stigma is displayed on the front of the column. In Orchis mascula the single anther may be seen just above the stigma. When it opens, the pollen is disclosed in two sticky masses—pollinia—prolonged downwards into slender stalks—caudicles, little tails. The latter end in small discs, which are extremely sticky; these are concealed in a small pouch, the rostellum.

Now when an insect follows the alluring fragrance of the orchid to its source, he alights on the labellum and advances towards the sweetness of the spur at its base. His mouthpiece inevitably touches the rostellum, which breaks at the contact. The sticky discs within at once adhere to the intruding part; the adhesive fluid in which they are bathed, when thus exposed to air, sets like cement within a few seconds. As the visitor withdraws his head, satisfied, he not only carries away the nourishment supplied by the labellum, but in token thereof wears a decorative pair of pollinia standing like two plumes erect on his beak.

"Now suppose (wrote Darwin) that the insect flies to another flower; it will be evident that the firmly attached pollinium will be simply pushed against or into its old position, namely into the anther-cell. How then can the flower be fertilized? This is effected by a beautiful contrivance; tough the viscid surface remains immovably attached, the apparently insignificant and minute disc of membrane to which the caudicle adheres is endowed with a remarkable power of contraction, which causes the pollinium to sweep through an angle of about ninety degrees, always in one direction, viz., towards the apex of the proboscis, in the course of thirty seconds on the average. After this movement, completed in an interval of time which would allow an insect to fly to another plant, it will be seen... that... the thick end of the pollinium now exactly strikes the stigmatic surface."

The beauty of the mechanism, its almost incredible fitness, is the precision of the movement of that minute disc; without which all the rest of the elaborate flower, its complex disposition, would be useless and without fruition.

Bonté vault mieux que beauté—bounty is worth more than beauty. Which is to say, flowering is of little value without fruit. Petals wither after a season, stamens shrivel; but if they have done their work, their place is taken by a pod or a berry, a fruit of some kind. They are as various, these fruits, as the flowers from which they come; some palatable, some poisonous, some succulent, some hard and dry, large and small, long and short, red and white and blue and green and black.

There is, I am told, a legal distinction between fruits and vegetables; doubtless as pregnant with ambiguity and confusion as most of the pronouncements of lawyers. Scientists also have wasted much breath on definitions—as if nature could be defined. In particular they have descended to distinctions between "true" and "false" fruits, which promise well enough until it is discovered that apples and strawberries are "false" fruits.

In truth all plants are vegetables, in the time-honored sense of that word and all parts of them, *including* their fruits; fruits, of course, are their edible products, whether of roots or of branches. Within recent years the good word vegetable seems to have lost meaning and by general consent fruits are now understood to be the products of flowers: plums, peaches, pears, melons, blueberries, breadfruit, coconuts, mangos, bananas, to the exclusion of potatoes, parsnips, carrots and lettuce-heads (but a cauliflower is undoubtedly floral).

The botanist finds it convenient to extend the popular concept of fruit to include anything of like origin, whether edible or not and whether served for dessert or with the meat. String beans in this sense are fruit, also a vast multitude of quite inedible pods, capsules, grains and nuts. Further than this it is not necessary to indulge professional idiosyncracies, nor to worry about the "falseness" of our desserts. Quite simply, fruits are the products of flowers. If the law decrees that a peach is a fruit but a bean is not, then—quite simply—the law utters nonsense.

A tulip, if allowed to wither naturally, is in due course replaced by a long green pod, in which may be found several hundreds of seeds.

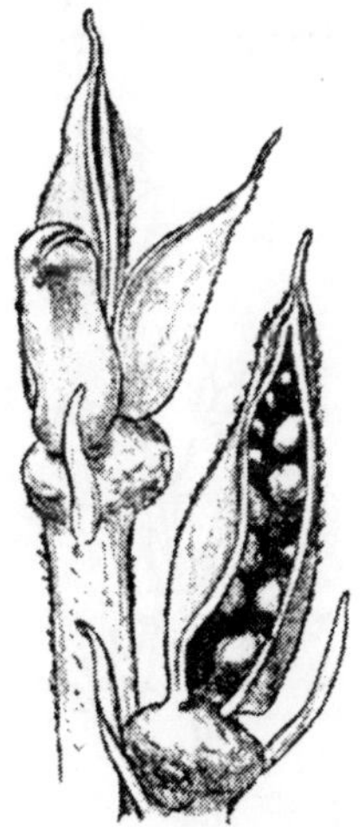

Fig. Pods

This capsule has developed from the ovary, the basal part of the pistil; the seeds—arranged in six rows within the three cavities-are the matured ovules of that body. The act of pollination, the arrival of pollen grains upon the stigma of the pistil, has several results. Not only does it make possible the development of the seed (in a way to be discussed in a later place), but it stimulates the growth of the ovary.

In all probability this is accomplished through the agency of certain growth substances derived from the pollen; at least it has been successfully imitated by the application of such substances, *in the absence of pollen*. In some species pollen will cause the fruit to develop without effecting fertilization; its two functions are apparently quite independent. In orchards it is well known that certain varieties will not set fruit unless pollen is supplied by other varieties.

The structure seen on a tulip-stalk after the flower has withered is a simple type of fruit: a matured ovary which has become a pod. Ofttimes the ovary becomes juicy, succulent, as it grows and becomes a berry; a grape is just that. Or it may change into a hard shell which encloses the seed or seeds: an acorn is such a fruit. A technical term is useful to specify the result of the maturation of the ovary and to avoid shocking those to whom it seems odd to call a grain of wheat a fruit: we name the developed ovary a pericarp.

Fig. Delphinium Flower and Fruit

Not only is it easier to speak of the pod of a tulip or of a peaplant by such an inclusive term;—we are also thereby relieved of the necessity for slandering an apple or a strawberry. Pods of lilies, peas, beans, poppies, larkspurs, geraniums; buckwheat grains and corn grains and oats; maple and ash keys, beechnuts, grapes and peaches and plums and cherries and raspberries, oranges and lemons,—all these are pericarps, varying in form, in texture, in flavour and in colour, but not in origin. The nature of some of our commonest and best-loved fruits, however, is a different story.

Consider with the eye of a naturalist the queen of flowers. Disregard for a time the exquisite imbricated petals and the five green recurved sepals which support them. Beneath all these may be seen a sort of urn or goblet, swelling up from the flowerstalk. On its rim the perianth, the throng of stamens are attached. As the petals wither and fall, the urn beneath grows larger, becomes bright orange or red and more or less succulent—a temptation at least to birds. This is the "hip" of the rose, the fruit which succeeds the flower; obviously it is not a pericarp, does not originate in the ovary of a pistil.

If a flower is slit lengthwise, cut through stalk and urn and calyx and corolla into two halves, its structure becomes clear. The inside of that cup is seen and its contents. The perianth and the stamens are, as was said, perched on its rim; the inside is lined with pistils.

Fig. Rose Buds

Each pistil has its own ovary, style and stigma—all the stigmas project in an almost solid mass through the opening above. After fertilization, each ovary becomes a pericarp; small, rather hard, bony.

These would without hesitation be called the seeds by most persons; actually each is a one-seeded nut. Meanwhile the cup in which they are enclosed continues growth also and becomes bright-coloured and succulent. What one might mistake for a large red berry containing seeds is really a transformed flower-stalk containing hard pericarps. Each of the latter corresponds, *in its origin*, to a pea pod or to a grape.

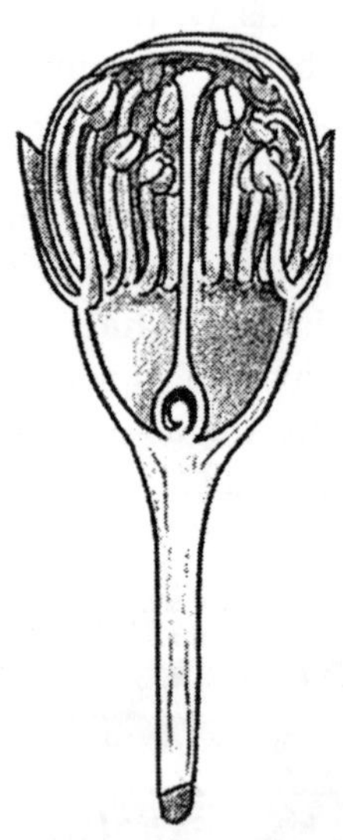

Fig. Cherry Buds

Examine next a cherry blossom. The sepals and petals are five. The stamens are many. Perianth and stamens arise from the rim of a deep cup, the end of the flower-stalk. The only obvious difference between this flower and a rose is that it has only one pistil, seated in the bottom of the cup. The fruit, however, which comes from the pistil, is quite different from either the nutlet of a rose or the pod of a tulip. The ovary enlarges to a really enormous degree, a thousand times its original size.

The outer layers become juicy, bright-coloured, succulent, delicious; the inner part of the same ovary becomes hard, stony. The stone of a plum or cherry is not the seed, though often so miscalled; it is part of the pericarp. Seed or seeds are inside, as anyone knows who has cracked open a stone. Almonds also are akin to cherries and plums, their fruit being dry and fibrous in its outer parts instead of juicy and edible. The inner part, the stone, is the nut which you can buy at the grocer's near Christmastide, with the edible seeds inside.

What of a strawberry? Here is a very succulent and eminently edible fruit, with the seeds on the outside. How can we reconcile such an arrangement with what has been said of the origin of fruits and seeds?

Fig. Cherry Ripe

We are betrayed here by the inadequacy of language to serve two masters. The two lords of thought are common usage and scientific need for precision. Either we coin fantastic and uncouth words for the need of science, or we strive to stretch common words to strange meanings. There is nothing in the strawberry flower or the strawberry itself that is incommensurable with the corresponding parts of cherries, roses, or tulips; but what we name the berry and think of as the fruit is not comparable to the pericarps of these other flowers.

Fig. Strawberry Flower-Stalk

It will be recalled that a rose has many pistils, each capable of giving rise to a pericarp. This is true of many flowers other than roses, which lack the urn which encloses the pistils. In a buttercup, a peony, a magnolia, the centre of the torus is crowned by a cluster of small pistils, enclosed by no cup other than the bright and transient blades of the corolla. This is true also of a strawberry flower.

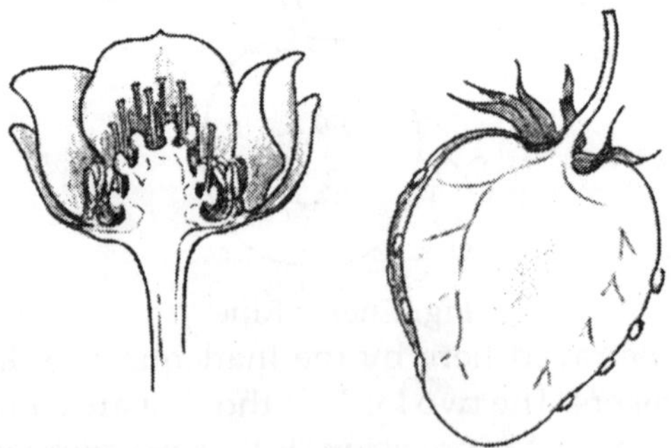

Fig. Seeds and Fruits of Strawberry

The tip of the flower-stalk, so far from being expanded into a hollow and cup-like container of the pistils which grow from its surface, projects as a minute dome in the centre of the *flower*, above the level of the perianth, carrying the pistils on its convex surface. One may speak of a strawberry flower-stalk as like that of a rose turned inside out; or vice versa.

The pistils are numerous and tiny; each formed of a round ovary from which arises, on one side near the base, a style tipped with a stigma. Each ovary contains an ovule, each stigma is daubed with pollen and each ovule becomes a seed. The ovary becomes not a large green pod nor an edible berry, but simply a thin hard shell immediately around the single seed. This is what the botanist names pericarp and others mistake for the seed. There are hundreds of them all over the projecting torus in the middle of the flower; it is the torus that enlarges, becomes red, sweet, edible. The botanist can delight his soul and shock our understanding by declaring that the strawberry "seeds" are "fruits" and that the strawberry itself is not a berry; —if we grant him his definition of a fruit as equivalent to the pericarp, the structure derived from an ovary.

Such is one of the chief causes of perplexity to beginners in science and one of the chief sources of pleasure to those who delight in curiosities: that things are not always what they seem. "Seeds" in the popular speech are usually small and hard and if planted generate new plants. A grain of wheat, the pericarp of a rose, the seed of a tomato—all these answer to the concept; but differ in structure and origin. Botanically seeds are certain specific tissues derived from ovules fertilized by pollen tubes; nothing not so formed may be considered a part of a seed.

Surrounding all seeds of flowering plants are always (at least for a time) those tissues which originate in the wall of the ovary. The source of confusion is the variety of forms and textures assumed by the pericarp. The edible parts of a grape or a banana or a tomato or a watermelon and both the flesh and the stone of a cherry—all these are pericarps. So are the pods of beans and peas, the capsules of tulips and lilies, the shells of acorns, the shell of a hickory nut, the husk and the shell of a coconut, the outer layers of a grain of wheat and the "seeds" of a strawberry.

Other members of the same great and various family which includes roses and strawberries and cherries are the blackberries, dewberries, raspberries and the like. The flowers in which these fruits have their inception are almost identical with those of strawberries. As development continues, after fertilization and the fall of the petals and withering of the stamens, the pistils (instead of the torus) enlarge and become the edible part.

These fruits are indeed pericarps; not single pericarps but clusters of them. Each individual segment, each delicious juicy globule of the entire fruit, is exactly comparable to a miniature cherry or plum.

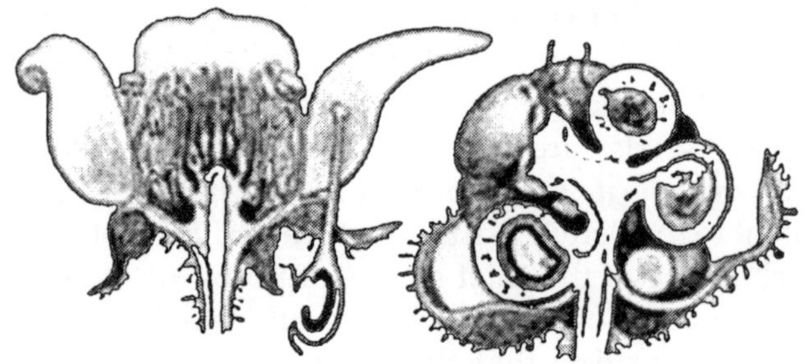

Fig. A Miniature Raspberry Plum

The inner layer, *around the seed*, is hard, composing a small stone; within is the seed itself. The stem-tip, torus, upon which all these small fruits are seated becomes fibrous, more or less inedible. That of a blackberry is detached as a part of the fruit; that of a raspberry remains on the flower-stalk when the fruit is pulled off.

In the same assemblage of plants belongs that noble fruit, the apple; with its relatives, pears, haws, quinces, medlars and sorbs. The best-known fact about an apple is also the clue to its botanical peculiarity. At one end, naturally, is the stem; at the other end, curiously, is the perianth, or the remains of it; apple-growers speak of the stem-end and the calyx-end of the fruit.

How does it happen that a fruit can intervene between stem and perianth? In lilies, cherries, strawberries, raspberries, the fruit, in any of its senses, grows from something in the middle of the flower. The other parts of the flower surround it and, as it enlarges, they remain near its base, near the stalk to which they are attached.

Cut an apple blossom lengthwise. Five sepals, five petals, a large number of stamens crown a stem-tip which is a cone with its pointed end towards the stalk. In the centre are five styles, partly joined near their bases—apparently with no ovaries. Look below, in the stem-tip, the torus; there, hidden in it, below all the other parts of the flower, are small cavities, five of them visible if you cut a flower crosswise. In these are the ovules. Here is then the compound ovary which belongs to the five styles. The pistil is complete, fivefold, but buried to its neck in the tissues of the stem which supports the flower. It seems as if the perianth and stamens were growing from the top of a large ovary; detailed study has shown that this is not really true. The ovary is surrounded and even blended with tissues comparable to the urn of a rose, the cup of a cherry blossom.

Fig. Seed Chambers

The fruit which grows from such a flower necessarily involves not only the pericarp, the part derived from the ovary, but also the surrounding tissues belonging to the tip of the stem and the bases of the perianth. Again we must digest a new meaning for that most versatile of words: fruit. Here it is neither pericarp nor any other part of the flower in the strict sense, nor is it only a swollen flower-stalk with a cluster of pericarps on its surface; it is a fusion of pericarp and other tissues into one indivisible mass. The edible part, "with a good savoure and merry," is a product of the end of the stem and the bases of the floral parts; the core contains the pericarp, part of which is a five-parted thin hard layer which limits the five seed-chambers.

The theme of fruit production is embellished with seemingly endless variation. The events which precede it, which make it possible, are surprisingly uniform. Pollen is carried to stigma, pollen tube penetrates style and ovary and ovule. It is instructive to cut through a young plum or cherry before the stone has hardened. The young seed, the former ovule, is filled with a jelly-like juice; at one end of this lies a small transparent capsule; in this in turn is a still more minute globule.

Fig. Seed Chambers of Apple

The last is the embryo, the future plum or cherry tree. In a mature seed it has greatly enlarged and consumed most of that delicate and nutritious jelly-like tissue around it and may be discovered as two flat white oval bodies joined at one end to a minute stalk. The paired bodies are the seed-leaves, the part to which they are attached is the stem, on the other end of which is the rudiment of a first root.

Roots, trunk, branches, leaves, blossoms,-all come from the slow and fortunate growth of this plantlet. A similar embryo may be found in the seed of a rose, of a strawberry, of a raspberry. The seed of a pea or a peanut is sufficiently familiar to every child: the two halves are the two seed leaves of the embryo.

Meanwhile around the seed occur all those changes which call into being the rose hip or the plum with its stone or the tomato or the blueberry or the blackberry or the strawberry or the acorn or the husk and shell of the walnut or the husk of the peanut; tremendous changes and enlargements of the ovary and frequently of many other parts of the flower.

The seeds are enclosed in layers of various hues, odors, tastes, golden, red, or blue, sweet, sour, or sickening, tart or mild; or in shells of varying hardness and differing intricacies of structure, with or without outer detachable fibrous husks. Even fruits which are apparently alike, seemingly duplicated by nature in several distinct experiments in the creation of families, may have originated quite differently. A grape is simply a pericarp, formed only from tissues of the ovary.

A blueberry is formed of tissues comparable to those in an apple, a pericarp embedded in other parts of the flower. An olive also is a pericarp, which like a cherry becomes partly fleshy and partly stony. A date pericarp is entirely fleshy; the stone within being the seed in truth. The history of a blackberry has already been set forth; it is a cluster of pericarps on one torus, from the ovaries of one flower. A mulberry, which so much resembles it when it is ripe, grows from a cluster of distinct flowers, each of which contributes a single pericarp to the whole fruit.

Fruits and seeds share with tubers and bulbs the doubtful honour of having made human agriculture and civilization possible. Here in handy packages are the most tasty and the most nutritious products of the industry of plants. It is after all of interest mainly to the naturalist that the deliciously flavored sweet succulence of a strawberry is contained in a stem-tip, that of a grape in a pericarp, that of an apple in a mixture of such tissues. With a bounteous hand Demeter offers us sugars in her cherries, pineapples, oranges, dates; oils and fats in olives, avocados, walnuts, coconuts, peanuts; starches and proteins and often sugars and fats in the vast tribe of cereal grains and vitamins withal and mineral salts and acids and esters to delight the palate.

All this wealth has meant a large contribution from the perpetual labour of the leaves; for all these good things are but the transformed products of that photosynthesis which was our first theme. So great is the drain upon a plant made by the development and removal of its fruit that the nutritive balance in its cells may be completely upset thereby. From having a surplus of carbohydrates, an apple tree may find itself with a relative deficiency of those primary foods; with the result that its course of development is changed and in its next season it forms many leaves but few flowers. All the fruit, all the grains, all the good wheaten flour and rice consumed over the earth represent the labour of green leaves.

In all this display of delectation plants seem to be actuated by concern for us and the birds rather than for their own good. Indeed many produce fruits in which are no seeds. Bananas, pineapples, many grapes, are of no conceivable use to the races which bear them except as bribes to man to continue to care for them; such races would largely die out in nature, or, if they survived, would do so by means of propagules other than their fruits.

These fruits, however, like the double flowers of our borders, are monstrous, in a sense artificial; they must not influence our botanical generalizations, our picture of the natural world. If we confine our thoughts to wild plants we see that their apparent benevolence is an illusion; as the production of bright petals results in something besides the gratification of man and the production of nectar has other effects than the nourishment of insects.

All these edible and tempting accessories have as their nearly inevitable result the transport of the seed, with its contained plantlet, to new ground, sowing it at random over an already crowded earth. Animals of various kinds make their meals of plums and cherries, hips and haws; ingest the whole fruit and digest the soft portions. The stone, or the hard pericarp, or the tough seed-coat itself, resists their digestive juices.

The seed thus enveloped passes out, is deposited again on the earth, with the precious embryo still living within: a strange parturition. Perchance the spot to which they have been translated contains no plumtrees nor rose-bushes. The advantage of being eaten is evident.

One of the most curious of plants is the mistletoe; remarkable enough to be sacred to our remote ancestors. A pale thing, always clinging to the branches of other plants, to elms and poplars and other trees. Its white berries still retain the power to dispense magic, to embolden timid hearts and bashful lips; a last relic of the veneration in which they were once held. They are filled with a very slimy flesh, embedded in which are the hard seeds. Birds feast on the berries; then are fain afterwards to clean their bills of the sticky, slime-enveloped seeds by wiping them on the branches of trees. So does this humble parasite get itself planted by helping to feed the birds.

The "dry fruits,"—pods, nuts, capsules and the like have a like use in the dissemination of species. The two halves of many a legume pod (belonging to vetches and other trailing weeds) are under tension, each kept extended and flat by the other to which it is joined. When the pod dries and the halves part company, they coil violently, throwing out the seeds within to some distance. Another common weed owes its name to a like habit: Noli-me-tangere, touch me not. Maple "keys" spread their veined wings on the air and the energy of their fall is partly converted into rotation. As they drift downwards, whirling, they may be wafted by any chance breeze to some distance from the parent tree.

The success of dandelions in migrating to fresh lawns needs no demonstration; it is due largely to the tuft of hairs which suspends the fruit in the air. Cockleburs and sandburs, sticktights and beggarticks are also fruits; they fasten themselves to passing animals and accompany them to their homes, where they may find space to germinate. The invasions of the flowering plants—quieter than those of men, often more persistent and more effective—have depended mainly on their fruits.

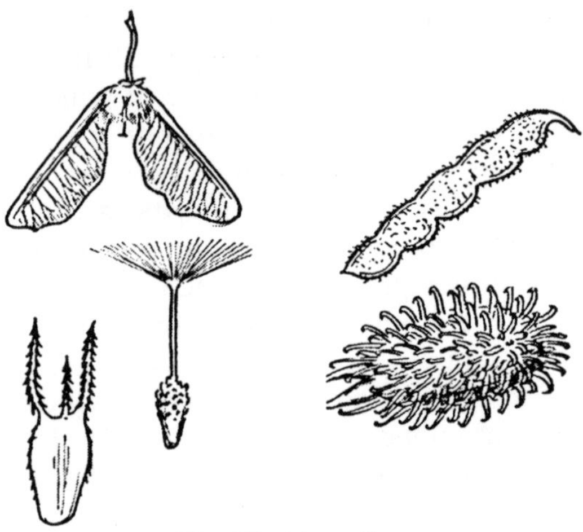

Fig. The Travellers

The first white travelers found in the northern parts of this green land a dense forest of tall evergreen trees, pines, spruces, hemlocks, firs: mile after mile of the straight neighborly trunks; shade-loving ferns and mosses and other herbs below in the semi-darkness, with the thrifty seedlings that would replace those of the giants that succumbed to wind or lightning or fungous decay.

An invitation to man's greed as well as a satisfaction of his wants; the tall stems were ruthlessly, unwisely, cut and carried away. A desolation was left behind; a sandy soil covered by the stumps of great trees, the burned remains of branches and saplings. In this devastated region gradually plants appeared, covering the scars, preparing the way for the next forest to grow.

The new plants were no seedlings of pine, fir and hemlock; but brush of various kinds, birch and aspen; bracken fern and raspberries. They were there, the little seeds carried by the birds in their annual flights, sown over the land, awaiting their moment.

Chapter 7

Peas in a Pod

The student of plants (and animals) finds himself in something the same predicament as Alice in the little dark shop on the other side of the looking-glass.

You will recall that, though the shop was crowded with a great variety of objects, these had a troublesome habit of being always on other shelves than the one that Alice was looking directly at. Living organisms, though we plant them, nourish them, harvest them, eat them, work them, ride them, are surprisingly difficult to pin down in our understanding; the more we learn about them the more difficult does it become to identify them.

A continuous stream of various sorts of atoms passes into and out of a living body. Cells die and disappear and new cells are fashioned in their places. Life is extremely difficult to characterize.

Even the boundaries of an organism are the less easy to determine the more one subjects it to scientific analysis. Only the relatively *non-living* parts are stable—the substance of bone and wood. A living being is less an exact structural unit than a sort of focus of activity, a whirlpool in the universal sea of atoms, which retains a continuity of form in spite of a constant flux of materials.

We remark the variation in the span of life of different creatures. A great tree may live for five thousand years: other plants and animals reckon their lives in months, weeks, even days. But the longest of these lives is but an instant in the history of the earth and the giant tree and the ephemeral fly are both the creatures of a moment in comparison with the slow swing of oceans, continents and mountain ranges. *Debemur mortis nos nostraque.*

In such a world, how is a science of biology possible? How can we found exact conclusions on such a phantom crowd of shifting shadows? The answer emerges from the facts detailed in the preceding chapter: living things reproduce. They form new living things in their own image, which carry the torch for a time and in turn relinquish it to their own offspring.

When we say "roses are red" we are thinking not of a particular garden and of the flowers there to be found but of an indefinite number of plants, past, present and future, all sharing certain characteristics which label them in our minds as "roses."

Every creature in the world represents a line of descent, in which there is recognizable a certain physiognomy, a certain stability; if we have no direct knowledge of this sequence of living individuals, we infer its existence.

Under one name, as Rosa rubra, we include thousands, hundreds of thousands, untold millions of plants, plants which grace a thousand gardens, plants which garlanded Roman nobles, plants which adorned the gardens of Al Mansur and of Elizabeth, plants which are not yet planted, which will grow in gardens not yet imagined. The name is the name of a species, that most characteristic of all biological concepts. Biology is the science of species.

We speak of the species with almost the assurance and affection that a chemist uses in speaking of a substance. True it is that it is a more or less arbitrary collection of impermanent aggregates of restless and invisible atoms. But we treat a species as something concrete, as if it were one vast and permanent plant or animal, only incidentally fragmented into individuals. Life is a necklace of many deaths strung upon a thread of immortality.

A famous satirist has pictured a land where machines had developed to such an extent that they threatened to take control of the men who designed them; not only to dominate their thoughts and so their life (as they do ours), but to form a permanent society of their own in which men would be merely the attendant slaves, the helots, the untouchables. The last and disastrous step in such an evolution (barely avoided in the imaginary country of the story) would be the production by the machines, without the aid of man or at least without his volition, of new machines to replace the old ones as they were worn out; to increase their numbers and to extend their dominion. The autonomy of man would be at an end when he saw the little locomotives playing round the door of the roundhouse.

Such a development seems so ludicrously remote that many persons—including biologists—have been led to assert that the power to reproduce is the ultimate distinction between the living and the non-living. This is an unfair attempt to substitute the unknowable for the unknown, to pretend that what does not happen is inconceivable and impossible. Certainly in this age the creation of new machines to replace the old, even of machines to kill men and destroy their works, has become a necessity against which enslaved man has not found the means to rebel. Is Samuel Butler's dream so fantastic?

Reproduction, in the biological sense, means the formation of new individuals in a precise pattern or sequence. Imagine that your automobile, a fairly complex mechanism with a definite set of characteristics, having retired into its garage, should detach from itself a rather shapeless and colourless and extremely small piece of substance and that this minute amorphous mass should then gather in necessary atoms from the deposit on the floor and walls and, enlarging greatly, fashion itself into a new automobile, complete even to useless bits of polished metal fastened on in the physiognomic pattern which distinguished its parent—or parents.

This highly improbable and doubtfully desirable train of events is indeed analogous to the reproduction of a living organism. A minute living unit is singled out, invested with individuality, enlarges eventually to a size comparable with that of its parent and assumes the particular pattern which has characterized a long line of its ancestors receding into the past.

Two questions may be distinguished and considered: First, how is the new individual formed? By what processes does the parent detach some part which can itself become a parent? Second, how does the new individual, once formed, take on the ancestral likeness? What impresses upon it the particular kind of life it is to lead, the particular course of development it is to undergo? Evidently these two questions are closely related; to answer one is to understand at least the meaning of the other. If we know exactly what a tulip contributes to the seed formed in its ovary and how, we have made the first long step towards understanding why the embryo will grow into a plant which will eventually bear a tulip and not a rose. To reproduce is to bequeath an inheritance.

The resemblance between parents and their progeny is generally referred to under the head of heredity. The first recorded comment on this perhaps regrettable state of affairs is, I suppose, in those lines of the Odyssey which say that "some children are like their fathers, most are worse, but a few are better."* Our concept of heredity is almost hopelessly tangled with our hopes and fears for posterity and with our personal vanities. If two unrelated boys have any features alike, it is a coincidence; but if father and son display the same passion for playing on the saxophone, it is accounted for by their physical relationship.

Here the trouble begins. Some one will protest that perhaps the taste of the father, his example and encouragement, have influenced the son during tender years; in a normal home he might never have developed such a failing. To proud parents who see in their offspring the aquiline nose or the love of fishing or the tendency towards absent-mindedness or the dislike of peaches which distinguishes themselves, heredity is a word which "explains" the marvel. To educators who lean towards another type of explanation, heredity is a myth to be discredited. So heredity is alternately made an excuse for the crimes of humanity and denied any share in human virtue.

It must be admitted, however, that there is some relation

between parent and offspring other than the purely educational or inspirational. Inquiry into the nature of this relation is as old as human curiosity. Why should a robin's egg become a robin rather than a blackbird or a blacksnake? It is certainly not merely because it is incubated by a robin, as cuckoos and cowbirds can testify. Why do frog's eggs grow so regularly into frogs and never into fish or waterweeds?

A multitude of different eggs may be reared in one aquarium or one puddle, under identical conditions and each grows unfailing into the sort of creature which gave it life. We are merely amused by medieval myths about seeds of trees which become butterflies or fish, of geese which become barnacles, of horsehairs which become snakes.

It is true that one still hears, even among educated persons who should know better, the question: "Do you believe in heredity?" But it turns out that to such persons "heredity" means the completely discredited legend of prenatal influence, the influence upon a child of events and thoughts experienced by the mother during pregnancy. After all, the excitement which attends a birth is not occasioned by curiosity about the species to which the new arrival will belong. Whatever be the truth about musical tastes and mental traits, a child is quite definitely characterizable as human by characters which cannot be ascribed to education or to the environment but which are certainly the result of something inherited.

If I seem to have wandered from plants, it is only that our concern with results of our own reproduction and heredity tinge our contemplation of these problems. Such a divarication is more apparent than real, for the facts of inheritance are surprisingly the same in both animal and vegetable realms. A group of ladies were needlessly shocked to learn that their children were little different in their mode of inheritance from so many peas in a pod.

The truth does not always flatter. Precisely what does an embryo in a seed inherit? To answer this we must describe in somewhat greater detail the events which lead to its formation.

The story begins with that small rounded body in the base of the pistil called the ovule, the "little egg." Small as it is, this body has a rather complex structure of many cells. If we disregard all but the essential detail, we may focus our attention upon one cell in the interior; this we may speak of as the "egg cell." The pollen deposited on the stigma has formed its long and slender tube, which penetrates the style and enters the ovary.

Once within the ovary it turns aside towards an ovule (doubtless in response to some substance formed by the latter) and grows into it, eventually pushing its tip between and into the interior cells which immediately surround the egg cell. The tip of the tube finally bursts in the ovule and empties into it the protoplasm which was within it. This consists of several things, among which are two small cells. One of these rapidly approaches the egg cell and enters it; the two become one cell. Likewise their nuclei come together, fuse, become a single nucleus. This new cell, thus formed from two parental cells, is the beginning of the new individual.

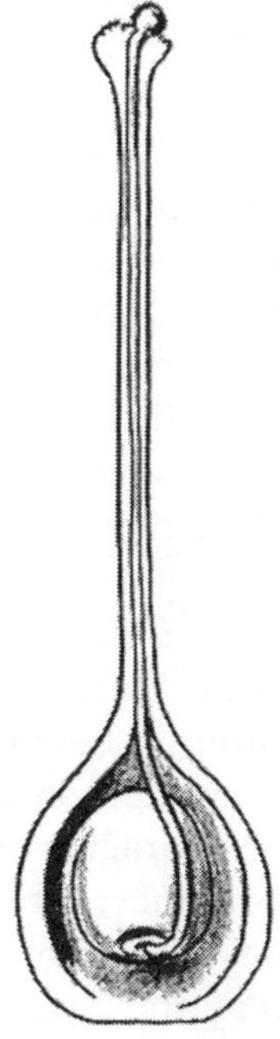

Fig. The Embryo Continues Growth

Among the numerous methods employed by plants to perpetuate their species, that which is here outlined is typical of a process very general throughout the plant world: the union of two single cells to make a new cell. The cells which unite are designated as GAMETES; the result of their union is a Zygote.

Almost all plants (and animals) form gametes which unite to form zygotes; every zygote is a potential new individual. The methods by which these cells are formed and brought together, the details of their structure, appearance and activity, are innumerable; they are fortunately not necessary to an understanding of the general behaviour of them all. Here—at the union of gametes—is the precise moment of origin of a new living organism, a new individual plant and here are the precise contributions of parent to offspring, of old generation to new. The young plant, which is still but a single cell, the zygote, has inherited the contents of two cells provided by its parent or parents, one from the ovule, the other from the pollen and *this is all*. It has, in fact, inherited *what it is*; its inheritance, at the beginning of its existence as an individual, is all its substance.

Growth begins. The single cell, by division, becomes two, the two become four and soon an embryo of many cells has appeared. Around it various food-containing and protecting layers develop at the same time from the tissues of the ovule which surrounded the egg cell. The embryo continues growth until it is recognizable as a small plant, with it first leaf or leaves attached to a short stem and the apex of the first root at the other end.

During this period it receives sustenance from the surrounding layers, but no further contributions of living substance. These later parental bequests are purely nutritive in their function; it has been demonstrated that they have no effect upon the particular character of the embryo.

After a period of dormancy, which may vary from a few weeks to many years, it resumes growth, breaks the seed coat, escapes from its envelop by the routine described previously. At appropriate moments in its history it forms leaves, branches, flowers, which bear the imprint of the race. They have those special peculiarities which enable us to recognize the plant as a lily or a rose or a grass—whatever its parent was.

Thus do we encounter the second problem of reproduction, the problem of heredity: what was in the single cell, the zygote, that made for the tulipy of the tulip, the roseness of the rose? Something there assuredly was, some quality in that inherited protoplasm different from the inherited protoplasm of other species. Yet in that first cell which contained all the inheritance no characters of the adult plant can be recognized. No flowers, red or white or otherwise, no leaves, no stems or roots; nothing but undifferentiated protoplasm, one nucleus with its surrounding cytoplasm.

What lay concealed in that indistinguishable lump of viscous colloids? The first clue to the elements of the situation were obtained, strangely enough, not by studying the cells with powerful microscopes, but by growing a number of races of peas in the little fenced garden of a small Bohemian monastery. The world now knows the story of the monk Gregor Mendel, whose name has been immortalized in the system of inheritance which he discovered. He was a modest person whom the great biologists of his day ignored. His training was in mathematics and the natural sciences.

His little leisure he resolved to devote to growing different races of plants and crossing them, in an effort to solve the old and baffling problems of the distribution of characters in breeding. He selected several races of the ordinary garden pea which differed only in minor ways; he grew them for several years to demonstrate that they were true-breeding and he crossed them by placing pollen from one upon the stigmas of another. What he found was as simple as it was unexpected.

If two parental plants differed only in stature, one being a tallish vine, the other a low bush (under the same nutritional and climatic conditions), the hybrids obtained by crossing them were all alike and all resembled the taller parent. *No plants were intermediate,* as most botanists of that day (and some of this) would have expected. In fact, so far as the visible results were concerned, the smaller parent had no hand in the production of the new generation. It made no difference which parent contributed the pollen and which the egg: it was as if the strain of heredity that makes for shortness, though it operated in the true-breeding short race, was completely lost when it came in contact with hereditary tallness.

At this point Mendel must have asked himself a very simple and acute question: Will this new generation, which is the same in appearance as the taller parental race, breed true, as that race does, when fertilized with its own pollen? It was easy to put this to the test. The answer was again simple, again surprising. When the hybrids were self-pollinated the resulting family of plants included *both tall and dwarf plants,* but again no intermediates; some looked like the one grandparental race, some like the other and these two types were very nearly *in a definite ratio of three of the taller to one of the shorter.*

Exactly similar results were obtained by crossing parents differing in other single characters, such as seed colour, pod shape, position of flowers on stalk and then self-pollinating the hybrids. The first generation exactly resembled one of the parents and the second contained approximately three-fourths of this type and one-fourth of the type that had apparently disappeared in the cross.

The explanation? Mendel's training in mathematics suggested it, for the ratio 3:1 is familiar in the binomial theorem. Let us suppose (he must have thought) that this particular character, tallness or shortness, is conditiond by a single unit in the inheritance (Mendel lived in the middle of the nineteenth century, knew little of cells and nothing of their methods of propagation and wrote rather confusingly in terms of the "characters" as inherited rather than something which underlay their development).

In the first hybrid, both units are brought together; one from the tall parent, the other from the short; the offspring inherits both tallness and shortness. Only one kind of unit operates; this we may call the "dominant," the other, which lies latent, the "recessive." Now suppose further that in the hybrid, when it in turn comes to reproduce, the two units which it bears separate in such a way that half its reproductive cells receive the dominant unit (which tends to cause tallness), the other half the recessive (which makes for shortness).

If we assign symbols to these units, the following diagram will represent what he supposed to occur: A hybrid, then, forms two kinds of pollen, here designated as T and S, in equal numbers; it forms also two kinds of egg cells, also labeled T and S, also in equal numbers. The supposed segregation of the two kinds of hereditary units occurs in the formation of *all* the reproductive cells, irrespective of their nature. When the plants are self-pollinated, we have no reason to assume that a pollen tube which contains T should be able to hunt out an egg which also contains T, or any particular egg whatever; as far as we know the behaviour of the pollen tubes is independent of these units within them and the particular combinations which occur are selected purely by chance. There are obviously four possibilities:

T pollen can fertilize T egg, forming a TT zygote.
T pollen can fertilize S egg, forming a TS zygote.
S pollen can fertilize T egg, forming a TS zygote.
S pollen can fertilize S egg, forming an SS zygote.

Any one of these combinations should occur about as often as any other, if nothing but chance dictates which pollen tube enters any ovule. Of every four new individuals formed by the union of gametes, in general one should have the hereditary constitution TT, two TS and one SS. But because of the dominance of T over S, the first three of these will look alike in size; only the fourth will betray the presence of the S units. Our suppositions about hereditary units exactly match the results obtained by experiment.

Since the hypothesis was created precisely for that purpose, this is perhaps not surprising. But it is one thing to work out an explanation—on paper—and another to reach some conviction that it really represents nature. Proof must be obtained by further experimental tests. If a hybrid really forms two kinds of reproductive cells, which differ only in a single unit of inheritance and are present in equal numbers, one should be able to predict the results of various sorts of crosses which involve such a hybrid. If it is crossed with the parent characterized by the recessive trait, one may work out the theoretical ratio of tall and short in the resulting progeny: it should be 1:1.

Mendel performed such tests, the results of which exactly confirmed his hypothesis. Furthermore, it is possible to predict the results of more complicated crosses in which the parents differ in more than one "unit character."

If plants with smooth yellow peas are crossed with plants having wrinkled green peas, the first generation will have smooth yellow peas, the second will include about nine-sixteenths plants with smooth yellow peas, about one-sixteenth plants with wrinkled green peas, about three-sixteenth plants with smooth green peas and about three-sixteenths plants with wrinkled yellow peas. Mendel made these complex crosses also and obtained results which conformed to the necessities of the theory.

When a theory successfully predicts the future and does so unfailingly again and again, not only for peas but for other kinds of plants and guinea pigs and fruit flies and crossable organisms in general,—it compels respect. In the early years of this century, even without other evidence to clarify the situation, the Mendelian laws of heredity came to be very generally accepted, although scientists had no clear idea of how any such series of units could exist in the reproductive cells and be separated in the way demanded by the theory. The hereditary units received various names; of which the name most in favour today is Genes. Opponents of the hypothesis also were not lacking and discussion at times became so lively as to employ invective as well as evidence. The findings of another branch of biology came to the rescue. Microscopic study had revealed the existence in cells of minute bodies whose behaviour was now shown to fulfill in every way the exacting requirements of the genes. These bodies are named Chromosomes.

It becomes necessary to redeem the pledge made in an earlier chapter to consider in detail the structure of meristic cells and the process by which they divide. The nucleus is now to be seen no longer as a minute sphere or hemisphere of colloids only slightly "thicker" than those of the surrounding cytoplasm. It is more like a rather loose skein of short lengths of yarn, the strands being extremely flexible, somewhat sticky, rather fragile, contractile and variable in length, thickness and position, but on the whole maintaining their form and other peculiarities through the life of the nucleus with great constancy; all bathed in a fluid and undifferentiated medium contained within the bounding membrane of the nucleus.

These strands are the chromosomes. So variable is their appearance and so complex their behaviour that for many years they were regarded with considerable incredulity. We now know (dissenting voices are few and negligible) that the chromosome-set is as constant a feature of a particular plant as the shape of its leaves and the colour of its flowers—indeed more so. In each of its myriad nuclei can be found just the same number of chromosomes—a number that is likely to be somewhere between ten and fifty-and, furthermore, just the same shapes and sizes of chromosomes if we compare like stages in their development.

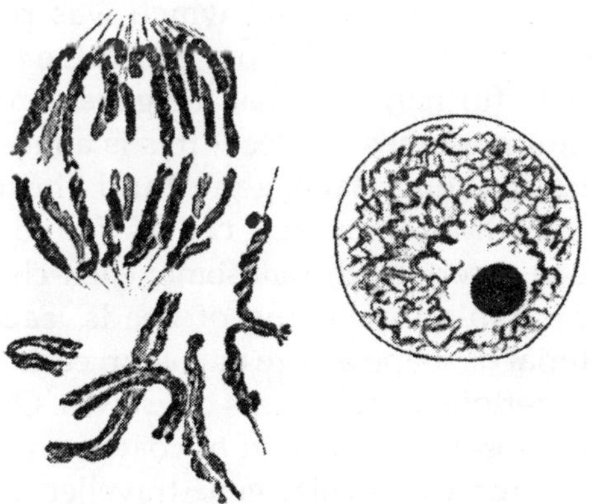

Fig. Nucleus and Chromosomes

If one chromosome is long and thin and has a lump at one end, its exact counterpart in length and caliber and even in its terminal adornment may be recognized in every cell of that plant. Still more important, the chromosomes are uniform *in different plants of one species*. They are permanent bodies, with individual characters of their own, which are perpetuated by cell division and endure through reproduction. In them reside the genes which are so important in the development of particular characters. In short, it is because of the chromosomes and their constancy in reproduction and growth that related plants develop like characters.

In the process by which cells ordinarily divide, each of the many chromosomes, seemingly tangled within the nucleus of each cell, becomes split or duplicated so that it consists of two like threads lying in close partnership. These two apparently undergo a mutual repulsion of some kind, or perhaps are dragged apart by some external force; in any case, they are shortly to be found at opposite ends of the cell, with all their fellows. Each group of half-chromosomes here becomes organized into a new nucleus.

A cell wall is deposited between these and we see two cells occupying the space of the original one, each possessed of the complete developmental mechanism which was the unique feature of the old one. This continues indefinitely, division after division, so that each cell of that plant possesses a replica of the chromosome-set which was present in the parent-cell of them all, the cell from which the plant grew.

It is known, furthermore, that the genes occupy definite spots in the chromosomes—indeed, this is almost all that we do know about them. Each gene (and there are many hundreds, even thousands, in a cell) is found always at a certain point in a certain chromosome; each chromosome is roughly comparable to a string of beads, each bead of a different material or shape or size or colour, each a gene which controls some definite departments of growth. Or we may say the set of chromosomes is a train of coaches in which every seat is reserved for a particular gene-traveller. A remarkable achievement in science, this discovery!—for no one has yet seen a gene and quite conceivably no one ever will.

The means by which we have learned not only that a certain gene controls some character but that it lies in a certain short region of a particular recognizable chromosome;—this story is one of the most fascinating chapters of modern science and the final paragraphs are not yet written. It has been accomplished largely by the development of great skill in the staining and preparation of chromosomes for microscopic study, so that every detail of these minute threads became as familiar to the cytologist (the student of cells) as the streets and houses around your home are to you.

Then it was found possible to cause certain abnormalities and injuries to the chromosomes by subjecting living cells to the influence of X-rays and other energies; the changes in the chromosomes—which were often limited to particular portions—could be observed with the microscope. And their occurrence was correlated with experimental breeding, which yielded information about the behaviour of the genes and deviations in it that were due to the treatment.

If the structure of the chromosomes and their paramount importance in development is once grasped, it is easy to understand that the "set" of chromosomes which is found in any cell of a given organism is the key to the peculiarities of that organism. You are human and your dog is canine largely and primarily because you have in every one of your cells a certain set of chromosomes which carry a human set of genes; while your faithful companion has also in every cell a set of chromosomes which differ from yours in thousands of genes as well as in visible microscopic details.

The genes have meddled in every process of growth of any living being, in the formation of its first undifferentiated cells, in the development of particular patterns of tissues, in the manufacture of particular pigments in skin or hair or petals, in the formation of enzymes which determine the chemistry of that particular chemical machine, in every detail of the manifold activities which we call living.

Notice particularly a rather odd thing about the genes, which was not grasped by some early students of heredity: *they do not explain differentiation*. That is, they do not enable us to understand why certain cells of a plant should become a root, others a stem, others a leaf; why the cells of a petal should adorn themselves with pink or blue, while the cells of a leaf put on their useful green. All the cells of one plant have the same genes—of that we are fairly certain; yet they accomplish a great variety of different things as the development of the organism reaches different points, each lending a hand in some particular phase, unconcerned in another phase.

But the genes do, quite definitely, help us to understand why the organism, *considered as a whole,* develops certain characters; why it develops roots and stems and leaves and has pink or blue flowers. Its set of genes, contained in its set of chromosomes, is master of the situation. And that knowledge at once provides the key to the resemblances which are due to inheritance. *The set of genes with which an organism starts life is derived from its parent intact.*

This is most plainly to be seen if we consider plants which are propagated by the artful care of the horticulturist or the farmer, without benefit of flowers. Everyone knows that potato "seed" is not derived from the potato blossom, but is a tuber cut up into pieces. The tuber is simply a branch of the stem, a branch which grows underground. It has much the structure of a stem, considerably swollen and full of food; it has meristic regions comparable to those on an ordinary stem. Each of the "eyes" is formed of a rudimentary leaf, a leaf that never attained its proper growth and, in the axil of that leaf, buds.

The buds in the eyes sprout into new growth when the surroundings provide the requisite stimuli. If meanwhile they have become detached from the parent tuber, from the parent plant, by the knife of the farmer intent on a new crop and set out in a new field, then the new growth becomes a new potato plant, standing alone and self-sufficient: reproduction has occurred.

Through all the formation of new parts the new cells arise by the process described above; the fact that the tuber, or a part of it, was removed from the parent has no effect upon the behaviour of the chromosomes in the dividing cells. Each new cell of the new individual thus formed has a set of chromosomes identical with those of the parent which formed the tuber. Its genes are the same as those of the preceding generation and in bodily characters also the resemblance is extremely close—providing that both grow under approximately the same conditions. Many are the tricks that can be employed to obtain new plants for the field, orchard, or garden. All depend on two peculiarities of plants: First, the presence in plants of so many undifferentiated cells; cells which have not matured in the fullest sense, but retain a sort of eternal youth which enables them, if proper stimuli are applied, to enter again upon a cycle of division, to form new cells, to develop new tissues, new parts, new organisms. Second, the occurrence in plants of persistent meristems, chiefly at the apices of stems and roots and in the axils of leaves, but also in other places. As a stem grows, as it forms new cells, not all of these become mature; some repeat the process of division and so on, ad infinitum.

The facts of plant propagation are well known. Branches are cut from geraniums and roses and chrysanthemums and hundreds of other kinds of plants and placed in moist sand to grow roots of their own and become complete individuals. Strawberries send forth their runners to root, later to wither or be cut off so that the rooted bud becomes a separate plant. Pieces of sweet potatoes (which are roots) or sugar cane (stem) are cut up and placed in the soil as if they were indeed seeds.

Certain plants are reproduced from their leaves. In the notches of the leaf of the succulent Bryophyllum are complete embryo plants, embedded in the ordinary tissues of the leaf; they await only contact with the soil to send forth their roots and new buds and some do not even wait for that, but may be seen dangling, a living fringe, around the margin of a leaf which is still attached to the parental stem.

It is well known also that the inheritance of plants produced by such methods is remarkably constant. It is a matter of guesswork what will come from a rose seed—and most of the guesses are apt to be wrong; but by taking cuttings you may grow new specimens which will reproduce in every detail the beauty which you wish to perpetuate. This has led some botanists to assert rather fantastically that new individuals produced by propagation without flowers are not really new individuals at all, but simply detached fragments of a single individual.

It is not worth while here to argue whether or not all the Burbank potatoes on earth or in it are parts of one vast plant. It is obvious that such an argument can serve only for the delectation of a speculative brain and can have no real meaning for science. However, the seed from which this monstrosity has sprung is the same which, in fairer soil, yields us a striking confirmation of our concept of inheritance. The plants propagated from stems, leaves and roots are as like as the stems, leaves and roots of one plant and all their cells contain the same chromosomes and the same genes.

When we examine into the minuter details of reproduction as it occurs in a flower, we find another kettle of fish altogether. We must begin by amending a certain statement made above. A cell of a flower or of a stem-tip or of a potato contains not one set of chromosomes but two. There are two samples of each chromosome and of every gene at every spot in each chromosome. The chromosomes are in pairs, the entire set is paired; the cells are, as we say, double or diploid in chromosome content.

This is true also of your own cells and of mine. The ordinary process of cell division in any organism perpetuates this doubleness. But when pollen is formed, a modification of cell division is set in motion. An unseen hand shuffles out this double set of threads into two single sets, one set in each of the daughter cells; the daughter cells and the pollen which comes from them, are haploid, single in chromosome content. An exactly similar phenomenon occurs in the formation of the egg cells in the ovules; they also are haploid.

But how, you may ask, does the number of chromosomes in a cell, whether a single set or a double set, concern us? What difference whether a flower is red because of one gene or two? As a matter of fact the so-called "reduction division" indicated above makes no difference at all in the breeding of plants and animals so long as the two sets of chromosomes in a cell are absolutely identical. To put it concretely, suppose that such a cell has a pair of genes which control flower colour, making it normally red; let them be designated by the letter A.

We write the "formula" of such a cell as AA, since it has

two identical chromosomes which carry at corresponding spots the two identical genes. In the formation of pollen and egg cells the sets of chromosomes are separated by the reduction division. One pollen grain, one egg cell gets only a single set, containing only once the chromosome which bears A; the formula of each haploid cell is A. When a zygote is formed by the union of gametes, the diploid condition is again established and may be written AA. The result is exactly as if the new plant had been formed by ordinary cell division like a potato sprout, without reduction and without union.

But when plants of slightly different genetic constitution are mated, many and curious are the chromosomal results. If, for example, a red-flowered plant is pollinated by a whiteflowered plant, the egg cell, containing gene A, is united with a cell from the pollen which contains no such gene, but instead, at the corresponding spot in the corresponding chromosome, another gene which we may label B; this gene causes the flowers to be white instead of red.

The hybrid plant is hybrid in every cell. Its formula may be written AB, which means that in each of its cells it has a pair of chromosomes which are not quite alike; one has gene A, the other gene B instead (in the corresponding spot). In other respects the chromosome set which came from one parent may be identical with that which came from the other, since we crossed closely related plants differing only in one character.

The hybrid, as Mendel found, may not show visible evidence of its mixed parentage; it may have red flowers, if gene A is dominant over gene B, the recessive. But when it forms its reproductive cells by reduction division, when the pollen, for instance, is formed in the stamens, it is at once evident that the situation is not so simple as that outlined above. If the double set of chromosomes is sorted out into two single sets, one of these must contain an A but no B, the other a B without an A.

The pollen must be of two kinds, half containing an A-chromosome, the other half a B-chromosome. The same is true of the egg cells. But this is exactly the gene-behaviour that was postulated by Mendel to explain his experimental results. The chromosomes actually may be seen, microscopically, to behave as he supposed that the hereditary units must behave. They are the constant bodies carried from generation to generation which control the development of specific characters and the reduction division is the mechanism which provides the various hereditary combinations which Mendel discovered by breeding peas.

As a result of knowledge gained with the microscope, carefully correlated with that gained by innumerable experiments in breeding plants and animals, a new science has won recognition in a world already well filled with sciences: the science of genetics. Its fundamental tenets are those outlined in the preceding paragraphs. Their starting point is still the hypothesis stated by a humble priest in an obscure monastery of central Europe, published in a local scholarly journal and disregarded by the intellectual world until after his death.

It is a remarkable tribute to the clarity of his thought that not only could he devise experiments that checked and verified his theory, but that subsequent research in a different field should also confirm it and that unseen and almost unimaginable units whose existence was demanded by his hypothesis should later be found to reside in clearly visible and well understood microscopic bodies in living cells, which behave in the way prescribed by him. Well has it been said that scientific progress consists of the verification of facts by means of hypotheses!

It is a matter of note that, even in this world of constant invention and discovery, in which we handle not only genes but molecules, atoms and electrons with the greatest of skill in our agriculture and industry, there are not wanting those who think that we can really believe only in what we see. To these dull minds a gene is still a "purely hypothetical" body and lacks a demonstration of "reality." For the sake of consistency they should also deny the reality of the ethereal vibrations which become audible to them in the speakers of their radio-sets, they should doubt the existence of the air which they breathe but have never seen; they should even doubt their own existence, for certainly the transient images seen in reflecting surfaces are not the living creatures which harbor these thoughts and make these statements!

It were well to remember also that, perhaps in younger and less staid days, they may have seen some things which were not "really" there! Let us say, rather, that for practical, scientific purposes reality is to be (at least partly) discovered in experience. Genes are real bodies (or at least points) and have a real existence, for otherwise our experience of our world would be different from what it is. Mendel could not have obtained his experimental results, nor could the modern army of geneticists and microscopists have had their particular experiences.

How different the modern notion of heredity, which has developed in such mushroom fashion in recent years, from all ideas that preceded Mendel's time! Older thought on this subject was decidedly stagnant, unproductive. Even Charles Darwin, famous for his powers of sustained and accurate observation, stultified much of his thinking about the origin of species and other matters by a completely mistaken view of inheritance. He supposed, with most others of his time, that the parent in some way *determined* the characters of the offspring; that each generation made a definite contribution of hereditary material to the reproductive cells.

He even went into some detail to sketch out a theory which involved a multitude of hereditary units, carried through the blood-stream in animals, in some unspecified manner in plants. Units from all parts of the adult body reached the reproductive cells and established themselves there. When development occurred, they reformed the parts in which they had originated. The racial inheritance was *remade* in every generation.

This has been well named the "transmission" concept of heredity, as distinct from the modern scientific view of the constancy of germplasm from generation to generation. The chromosomes and their genes, be it noticed, are furnished to all the cells of a new individual, including his reproductive cells; the latter derive their inheritance (which they in turn are to pass on to the next generation) not from the cells of the body around them but *directly from the preceding generation.*

Recall the history of development of a plant from the embryo in the seed; is it not plain that the meristems at the tips of roots and stems, while they form mature tissues of all kinds, *are not formed by them.* They constitute an unbroken line of undifferentiated cells from the original germ of the individual to the germ cells of the flowers which they finally form. It is true that new plants may be formed from relatively mature cells of stems, roots, or leaves; but it is plain also that these cells receive no contributions from the remainder of the plant body, but develop by virtue of the common inheritance of which they have a share.

This puts a new complexion on the famous old controversy entitled "the inheritance of acquired characters." If Darwin were right, each individual, formed by contributions from all the parts of its parents, should reproduce their peculiarities with considerable exactitude. If a plant suffered an amputation or some other malformation or mutilation, the next generation should show the effect, since it would lack hereditary particles usually contributed by the part affected. Conversely the role of an overdeveloped organ would be accentuated.

In fact, the supposition that such things actually occur has dominated thought on heredity through most of history. So "natural" did it seem that such a mode of inheritance was not seriously questioned until recently and eminent biologists and philosophers from Aristotle to Herbert Spencer vitiated their biologic thought therewith.

A generation of children has read how the elephant's child acquired a long trunk, how the camel got his hump and the leopard his spots; "and from that day to this, Best Beloved." The American Indian accounted for the woodpecker's red head as a reward bestowed by a spirit upon an ancestral woodpecker. So well esconced is this way of thinking in our mental furniture, that even today we resist the evidence provided by science and consider it quite plausible that Manx cats are tailless because of systematic mutilation by their ancestral owners. Writers of books on birds speak of the songs which the singers have "learned" from their ancestors. Yet we should hesitate to arrange scholastic curricula on the assumption that if the parent has excelled in mathematics his son needs no instruction in it. Nor do we now believe that the unfortunate men who survive a war without their full tale of limbs have produced since then a number of incomplete offspring.

Those who have clung to belief in the inheritance of "acquired characters" have evaded such absurdities by supposing that what does not happen in one generation may in some way be operative over a period of years, providing the cause is constantly present; they speak of a character being "impressed" in time. Upon such a basis the Chevalier de Lamarck founded his theory of evolution. Naturally we cannot furnish conclusive evidence on the effects of causes which must operate through millions of years.

But certainly all modern attempts—and they have been numerous, sincere and ingenious—to substantiate by experiment the claim that the bodily history of the parent has a direct influence on the characters of the offspring,—all have failed. The question boiled up at the close of the last century, was the focus of bitter controversy and considerable experimentation; with negative results. The most famous investigation was that of the zoologist August Weismann, who systematically cut off the tails of mice and bred the mutilated animals, repeating the operation for nineteen successive generations. The preserved skins of these small martyrs to science still bear mute testimony to the inefficiency of such methods; the last being as well tailed as the original animals. Folklore and the tales of Rudyard Kipling have value of many kinds; but represent a profound biological error.

After all, what are acquired characters? Characters of the body acquired during its development? All normal characters are so acquired. Flower colour is present only in flowers; in the seed which you place in the garden is no colour. Hair colour develops in our own children as we watch them. Even at birth (and development has already been going on for some time) they commonly have little or no hair. My characters, my hair and eyes and teeth and arms and legs and strength and weakness and idiosyncrasies—these are *my* characters, developed by me; no one else ever had them, though my father and mother and sundry grandparents had characters something like them. To say that a child inherits his mother's eyes and his father's nose is not to be taken too literally. *All characters are acquired during life; none is inherited.*

Once this is grasped, much of the careless generalization which still unfortunately characterizes writing about heredity vanishes into smoke. It is to be regretted that so many people—even Doctors of Philosophy—feel called upon in their treatments of sociology, psychology and the like, to make pronouncements upon heredity which are obviously based on a lack of acquaintance with the biological truth. Once when I was a professor in a university and sat with other professors on a committee, a colleague, who was not a biologist, turned to me and said: "The trouble with the way you fellows teach biology is that you don't make it plain that structures are inherited but functions are not."

There is a certain dangerous plausibility about such a statement; as one might say: the structure of a certain kind of automobile is uniform because they are all made in the same factory and from. the same patterns, but their functions—what they do—depend on the men who drive them.

There is certainly truth in such a statement. But as certainly is it true that in no conceivable ownership does an automobile, without very extensive changes of structure, swim, fly, or play the piano. There is a certain necessary connection between structure and function, such that the latter is limited even more by structure than by circumstances. If we are to be charitable towards the biological ignorance of sociologists, we must interpret the statement quoted above in this sense: Resemblances in structure between parent and offspring are due to heredity, but their functions resemble each other only because they are limited by similar structures, not directly because of heredity. Careful thought reveals the untenability of even this broadly interpreted point of view.

What is a structure? Is a fingernail a structure? Certainly the nail now visible on your right forefinger was not in existence a few months ago, still less at birth. It has developed, grown and continually is replaced. In no real sense is that structure *inherited;* what was inherited was some very un-nail-like protoplasm in which was a gene or set of genes which had to do with the growth of nails.

Indeed, it were more proper to say the *function of growing fingernails* is inherited than the nails themselves, though even this is not quite accurate. Let it not be thought that I have chosen an unfair example; the fingernail situation is general. A leaf is certainly a structure and as certainly no leaf on the oak this year was inherited from the parent tree a hundred years ago, for every leaf was formed no longer ago than last summer.

The truth is that neither structures nor functions of mature organisms are inherited, but only a small bit of living protoplasm, which has a structure and set of functions of its own and in it structure and function are completely inseparable and both inherited. The sociological twaddle derives from a too childlike faith in the literal meaning of certain figures of speech, which themselves belong to the transmission myth of our folk-lore.

Let us not rush, with some overenthusiastic students of biology, to the other extreme. The environment is important and the inherited germ does not absolutely and finally "determine" what the mature plant or animal shall be. A question which has tormented many a student of the social sciences (if they be sciences) is: Which is more important, heredity or environment? It is natural enough that one who works with the distressed classes of humanity should see in the squalor and ignorance with which their children are surrounded a sufficient cause of ill health, bad habits, even criminal acts.

Those who believe that "structures only are inherited" conceive of heredity as having to do only with the physical resemblances among people, while the environment is entirely responsible for what they do. While the essential falsity of such a position should now be evident, it would betray an equal failure to grasp essential truth if we follow certain others who declare that crime, alcoholism, insanity in all its degrees and the entire social situation which such defects create,—all are to be assigned to heredity.

The truth obvious to any biologist is that neither heredity nor environment can by itself account for anything. If it takes two to make a quarrel, it certainly requires no less to initiate a physical or chemical reaction. The propulsion of a bullet by the explosion of gunpowder depends primarily upon the presence of certain chemical substances which compose the gunpowder. If sand or salt is substituted, the explosion does not occur and the bullet does not travel.

On the other hand, the gunpowder will not give the desired result unless it is contained in a suitably constructed chamber and subjected to a proper igniting spark. If it is ignited in the air, it may burn slowly instead of exploding and will propel no bullets. If the gun is improperly designed, it may be shattered by the explosion and the man who holds it may be killed instead of the squirrel on which he intended to sup.

So if elephant protoplasm could be substituted for rabbit protoplasm in an hypothetical organism at the threshold of life, the result of the processes of development (supposing them to occur at all) would be quite unusual for the rabbit world; even if the environment remained that in which rabbits usually make their appearance. We are reminded of the famous medieval recipe for the creation of mice. Merely place some old rags and bits of cheese in a drawer (it ran) and leave them undisturbed for a week or two and when you open the drawer again you will find that mice have been engendered. The credulous men of that day had such faith in the power of environment that they dispensed altogether with the living germ.

On the other hand, even rabbit protoplasm cannot develop in a void and the exact course of its development will be markedly influenced by the food, light, air and other environmental factors with which it comes in contact. This is well seen in plants. Plants grown indoors are frequently pale and spindling and we know that these characters are correlated with the weakness of the light in which they develop.

The same plant grown in brilliant sunshine and in a dry air will form not only a shorter and stouter stem and greener leaves, but its leaves will usually be smaller and thicker and more heavily coated with cutin. Plants which lack a sufficiency of iron in the soil will be pale, even yellow instead of green; a deficiency of phosphorus, on the other hand, will lend a certain reddish tinge to the margins of the leaves.

But to return to the question: Which is the more important, heredity or environment? Is heredity more important in animals, environment in plants? Actually it is impossible to give an answer, for the question itself has no meaning. The plant breeder can show you corn plants which are yellow instead of green; they have been grown in a normal soil under normal illumination but possess a certain genic constitution which makes them unable to develop chlorophyll. The plant physiologist can show you corn plants which are likewise yellow but because of a deficiency in the soil; their inheritance being entirely normal.

The environment was important in the development of *their* abnormality, as the inheritance was for the first. A plant, or an animal, is the result of a certain heredity operating in a certain environment. It is worth while considering the question only to point out that certain characters are easily modified by a change in the environment, while others cannot be so influenced by any experimental procedure. We can very easily cause changes in the colour of a Hydrangea flower or of a corn leaf; but by no conceivable laboratory technique can we make a Hydrangea flower grow in the place of a corn leaf.

A living body—we repeat—is not a very clearly defined thing. Its life involves the surrounding world continually and in a variety of very complex ways. The environment is a part of its life; life being a name which we give to a set of activities which depend upon materials and conditions both within and without the organism. As for the structures of the organism, they are in a very real and literal way the products of its functions, as its functions are at least to a large extent the results of its structure. Heredity and environment are inseparable in their effects upon development; every character bears the imprint of both. To ask which was most important in a man's growth is quite like asking which is most important in his everyday life; could he get along most easily without his structures or without his functions? Without his stomach or without his digestion?

From discoursing on the likeness of peas we have reached questions whose right to appear in a book on plants may be questioned. It is true, as I have remarked before, that in many respects (and this is one of them) human babies are much like peas in a pod and the problems encountered by a venturesome seedling as it thrusts out from the sheltering soil are not entirely dissociated from those which confront our own progeny. It is time to return to our vegetables and to their manner of inheriting.

The discoveries outlined above were so productive of new knowledge that it is no wonder that the collective head of biology was somewhat turned for a time. The success of the Mendelian experiments was due to the focusing of attention upon single characters, "unit characters." For a time organisms were rather absurdly referred to as "bundles of characters," with a lack of realization that characters are abstractions from the whole organism.

Any "unit" character that you name I can treat as two characters if I wish; the number of characters is infinite. Fortunately it was early discovered that some characters are conditioned by two distinct genes in two unrelated spots in different chromosomes; some by three or four. Attention was then transferred to the genes from their visible symptoms and it was realized that *they* were Mendel's units.

It is now known that many different genes may cooperate in the development of a character. The colour of the flowers may seem to be a unit in inheritance if the plants used in an experiment differ in only one of the genes concerned; in other experiments it may behave differently. The genes, however, always seem to behave as units in their distribution from generation to generation. Not only do different genes influence the same character, one gene may play a part in several different characters. We may sum up their behaviour by saying that all the genes in the set work together to produce the organism—development is one great complex reaction.

The genes have not yet been seen, although, in the manner briefly indicated above, the exact hiding-place of some of them has been identified. A student may peer through a microscope at a minute twisted thread and say, "There, in the short region between the end and that colourless spot, is the gene which causes this species to have red flowers. I know not what it is nor how it works, but I know it is there." It has often been suggested, with some cogency, that a gene is a centre of enzyme production. Certainly their vast effects are out of proportion to their minute quantity; this is true of enzymes also. It has always been difficult for the physiologist to conceive of something so evidently reactive as a gene being of the constancy in position and nature that has been demonstrated for them.

The knowledge now so rapidly growing of "growth-substances"—hormones, auxins, vitamins and the like—will perhaps lead us eventually to an understanding of genes; certainly the latter may be said to furnish yet another example of a principle enunciated on a previous page: that the activities of living protoplasm are characterized by their dependence upon very potent substances present in very minute quantities.

The information about genes presented in this discourse is limited to the simplest. If one pursues the subject further one is rapidly involved in a highly technical and complicated subject—in which great rewards still await the researcher. Even this elementary treatment, however, enables us to understand with the utmost clarity certain phenomena of inheritance which were puzzles to our fathers. For instance, notice how characters may often "skip a generation"; an event which was referred to as "atavism" and provoked elaborate and recondite speculation.

Now we see it simply as a normal result of dominance and recessiveness of genes. A tall and a dwarf plant are hybridized; the hybrid generation is all tall and bears no resemblance in this respect to the dwarf parent; in the second generation—providing sufficient individuals are raised—about one-fourth of the plants are dwarfish like one of their "grandparents." If only a small number of individuals is grown in each generation, the recessive character, purely through chance, may fail to appear at all. It may be hidden for many generations and its final emergence cause considerable surprise, being greeted as an almost miraculous example of resemblance to some forgotten ancestor.

The matter of "sufficient numbers" will bear scrutiny. It has been emphasized throughout this exposition that the combination of different kinds of gametes from pollen and ovule is purely a matter of chance. If we are following the history of a single pollen tube, we cannot prophesy with which of two types of egg cells its gamete will unite. But if we have the results of many such unions, we can say with some confidence that a certain proportion will be of one type and the remainder of the other.

In the same way, if you toss a coin it is impossible for you to say with absolute certainty which side will be visible after it comes to rest on a flat surface. But if you do it a hundred times, you can say without serious misgiving that *approximately* fifty times the coin will fall "heads" up, the rest "tails." Even if we do it a thousand times we can still not predict the outcome of any one throw and consequently we cannot predict the *exact* number of each alternative.

But the more often we make the trial, the nearer, we know, will the result approach a ration of 1:1 and this much is valuable information, especially in the breeding of plants and animals. It is evident that we can make very few assertions about the results of inheritance in a human family. But a corn plant can yield a family of several hundred and the general proportion of different types can be known beforehand if we know something of their parentage.

The practical breeders, the agriculturists, the horticulturists, have in fact used the Mendelian principles to great advantage in recent years in the development of new and better types of domestic plants and animals. The understanding of the mode of inheritance has made it possible, for instance, to combine the ability of certain kinds of wheat to resist the rust disease with the desirable milling qualities of the rust-susceptible races. Thus does an apparently academic experiment conducted in a garden in Bohemia come to have vital significance in the economics of North Dakota.

Other interesting facts come to light if we scrutinize the numerical relations involved. I have described the mode of inheritance of only a single pair of genes situated at corresponding spots on corresponding chromosomes; such genes may be spoken of as alleles. But it is possible for two crossable parents to differ in more than one way and for the hybrid to be of mixed descent not only in the alleles (let us say) A and B but for a distinct pair of alleles C and D.

The story is the same as for each of these pairs taken separately and it is not necessary to detail it here; but if we catalogue all the possible types of offspring obtainable we will naturally have a more complex population than if only one pair of alleles is involved. Reference has already been made to such populations obtained by Mendel. Peas which differed in both the colour and the surface of the seeds yielded (in the second generation) four types of offspring; the ratio of these was approximately 9:3:3:1.

If *three* pairs of alleles had been involved in the differences between the first parents, there would be in the second generation *eight* types of plants in the approximate ratio 27:9:9:9:3:3:3:1. If there had been *four* pairs of alleles *sixteen* types of offspring are obtainable and, in general, if there are *n* pairs of genes different in the parents, the number of possible types of offspring in the second generation is 2^n.

Since it is easily possible that cultivated plants such as roses and apples which are usually cross-pollinated are hybrid for, let us say, twentyfive pairs of genes, it is easy to see why it is impossible to make predictions about the outcome of seed from such plants: from such parentage there would be over *thirtythree-and-one-half millions* of possible types! It is fortunate indeed that other methods are at hand for the propagation of our choice flowers and fruits.

At the other extreme is the situation that results from the continued self-pollination of a race of plants. If we start with a single individual which has a single mixed pair of genes, AB and self-pollinate it, we get, as has already been shown, three types corresponding to the genic formulae AA, AB and BB in the ratio (approximately) 1:2:1. (The first two types may, because of dominance, be indistinguishable to the eye; but this is of no moment in the present discussion.)

If now we self-pollinate all these plants, one-half of them, AA and BB, will come true to type, each forming only one kind of germ cell; while the other half will repeat the behaviour of the first generation, forming the same three types in the same proportions. If we suppose that every self-pollinated parent forms forty new plants, then from four members of the first generation we could obtain offspring as follows: from the plant labeled AA, 40 new AA's; from the plant BB, 40 new BB's; from the 2 plants containing AB, a total of 80 plants of which 20 would be AA, 40 AB and 20 BB.

If we add like types, we have 60 AA, 40 AB, 60 BB, a ratio of 3:2:3. Compare this with the ratio of the same types in the previous generation, which was 1:2:1; in the former family the AB individuals were half of the total, in this generation they amount only to onefourth. The same reasoning continued into the following generation reveals a ratio of 7:2:7; in the next we find 15:2:15 and so on. In general, in the *nth* generation the ratio of the various types would be $2^{n''} - 1:2:2^{n''} - 1$.

This means that in the 25th generation we should have, instead of the single hybrid race with which we started, two purebreeding races (AA and BB), with a very minute chance-about one in seventeen millions—of an occasional hybrid (AB) showing up. The effect of continued self-pollination is to concentrate genes in the pure condition in pure-breeding races. Such races are found in species which habitually selfpollinate, such as beans and wheat.

Exactly the same effect can be obtained among animals by close inbreeding; i.e., by brother-to-sister matings. This biological situation underlies the almost universal taboo against such a scheme of mating in human society. For, unfortunately enough, it is true that there are in most human families certain undesirable genes which tend to produce mental or physical defects. These genes are almost all recessive, perfectly harmless so long as they are concealed by the dominant corresponding genes (alleles) in mixed matings.

But a series of brother-to-sister matings, or even inbreeding not quite so close as this, tends inevitably to concentrate these genes in the pure condition in certain lines of descent; with the appearance of numbers of defective and undesirable persons. Although this subject has been carelessly treated and overly exploited by certain writers—for insanity and feeble-mindedness are difficult subjects and rarely capable of clear analysis in hereditary terms—it is undoubtedly true that undue numbers of mental and physical defectives occur in isolated communities which are denied intercourse with other groups.

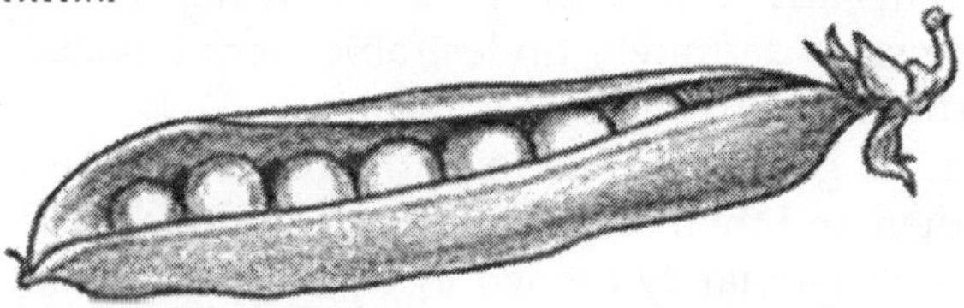

On the other hand, the same process can also concentrate in more fortunate lines of descent genes which make for exceptional talents and virtues. Families are known in which distinguished scientists were succeeded by others of like distinction; jurists, statesmen, musicians, have been similarly related and the succession is not to be entirely accounted for by education and example.

Something of the same occurred also in ancient Egypt in some of the reigning dynasties. Brother-and-sister marriages were common, were even encouraged for those of high birth and resulted in long lines of rulers of exceptional ability. There seems to be nothing intrinsically wrong in close inbreeding, only as there happens to be something intrinsically wrong in the racial inheritance.

Chapter 8

Fern and Moss

Fern seed has long been celebrated for its remarkable powers. Its fortunate possessor is able to see buried treasure shine with a bluish light; or, if he throws some of the seed into the air, it will fall to earth where treasure lies hid. If he puts fern seed in his shoes he is rendered invisible and can come and go on his affairs unseen by his fellow men. This property makes the seed definitely undesirable from a social point of view; fortunately it is extremely scarce.

It can be gathered only at midnight of midsummer eve, for only then and then only for a moment, do ferns bloom and the seed is immediately carried away by the fairies, to whom it is of obvious importance. Furthermore, even at the prescribed moment it may be obtained only by the use of certain rites and precautions, which it is not lawful to reveal here.

Unfortunately for those who hope for supernatural help in the realization of their dreams, the inquiring eyes of science have discovered that fern seed, far from being rare, is extraordinarily abundant and that it has apparently no magical virtues. It is true that, so far as I know, no botanist has been at pains to investigate the properties of fern seed gathered with appropriate incantations at midnight on midsummer eve; but we may be pardoned a certain skepticism about the value of such a procedure, particularly since we now know that fern seed is regularly produced without flowers at all.

However, this may be a worthy subject for some future candidate for the degree of Doctor of Philosophy and it may be proper for us to suspend final judgment on the processes employed by ferns for reproduction at the aforesaid moment. Meanwhile it is not without interest to contemplate the methods of reproduction that obtain during the rest of the year.

Fern-loving housewives have been known to take soapsuds and scrubbing-brush to remove the little brown scales or dots which sometimes appear on the lower surfaces of the leaves. If in their zeal they succeed, they can only injure the plant. For these are not insects nor abnormalities caused by some parasite; they are the normal reproductive structures of the fern.

Do but lay such a fern leaf on a sheet of white paper, the lower surface next the paper; over it place an inverted bowl or any cover to protect it from moving air. In a few hours remove the leaf: the paper will retain its outline marked in brown dust. This is fern seed—by the millions.

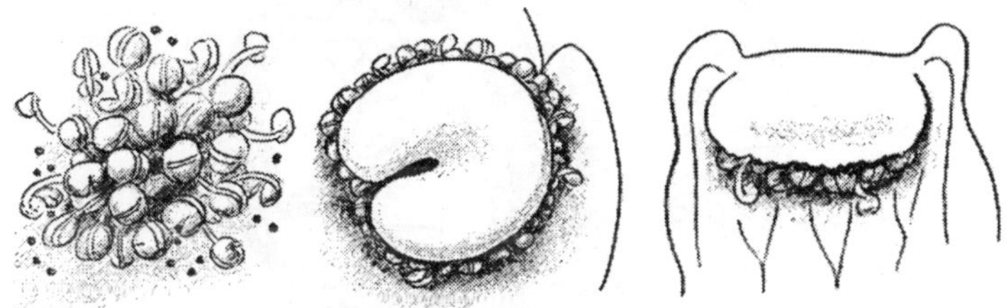

Fig. Under Fern Leaves

Fern seed is microscopic; so also are the structures in which it is formed. If the leaf is placed under a microscope with its lower surface up and one of the so-called "fruit dots" visible and magnified about twenty times, a wealth of beautiful detail is at once visible. Each dot is a cluster of minute globes which stand erect on hair-like stalks. In young stages these are green, later they become brown, finally transparent. Those of some ferns are protected by thin scales which grow from the leaf at one side of the cluster and arch over them, or rise on a stalk from their midst like an umbrella.

Or the group is at first covered in a transparent sac which splits open in the centre or at one side. In some ferns these reproductive structures stand near the edge of the leaf, covered by a membrane which extends inward from the margin.

Each small globe is a hollow sac, filled with brown dust,

with impalpable grains to be seen only with the microscope: fern seed. So minute are these grains and so simple in structure when compared to beans or mustard seed, that the botanist prefers to distinguish them by another term; they are called Spores (a word which means seeds, but, being Greek, is less liable to be understood in the usual sense). Each is but a single cell, an undifferentiated unit of protoplasm containing a single nucleus and surrounded by a rather thick and rigid wall; in it there is no plant in miniature embedded in a nutritive medium. Each measures only about 0.05 millimeter or less (1/500 inch).

The small globes which at first contain this stuff split open, yawn widely, often turning themselves inside out.

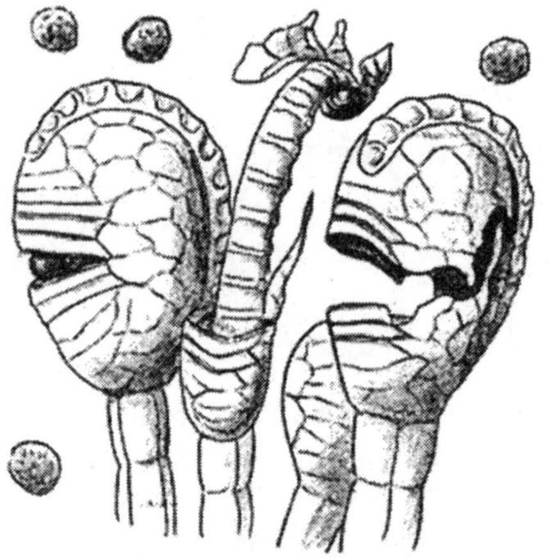

Fig. Fern Seed

They are made of transparent flat cells with curved walls; on one side, running about three-fourths of the way around from the base and over the top, is a crest of projecting cells whose cross walls are thickened and yellow. This ring, when it loses moisture in a dry atmosphere, straightens out with some force, tearing apart the more delicate cells to which it is attached and spilling the spores within.

Later the motion may be suddenly reversed; the spore-case may suddenly snap closed again, with such violence as to throw to some distance any remaining spores clinging to it.

Each spore-case may contain 64 spores; there may be

twenty cases in a single small group and three or four thousand of these groups on a single leaf. One small fern leaf may produce three or four million spores; a large plant may form twenty or thirty million spores—each a potential new plant. So far from fern "seed" being rare, it is a characteristic example of the prodigality of plant life. If every spore of an ordinary large fern lived and came to maturity, there would be, let us say, 25,000,000 new fern plants.

If we suppose that each spore had landed far enough from its neighbors so that all these plants might live and each new plant occupied about 1 square foot of land, this population, the progeny of a single parent, would cover about 600 acres. If the whole process were repeated once more, every one of these ferns again producing 25,000,000 new ferns, the resulting greenery *would cover the entire continents of Europe and Asia*! If we compare this theoretical world of ferns with their actual distribution over the earth, it is evident that the mortality of fern seed runs into incalculable billions.

They are made of transparent flat cells with curved walls; on one side, running about three-fourths of the way around from the base and over the top, is a crest of projecting cells whose cross walls are thickened and yellow. This ring, when it loses moisture in a dry atmosphere, straightens out with some force, tearing apart the more delicate cells to which it is attached and spilling the spores within. Later the motion may be suddenly reversed; the spore-case may suddenly snap closed again, with such violence as to throw to some distance any remaining spores clinging to it.

Each spore-case may contain 64 spores; there may be twenty cases in a single small group and three or four thousand of these groups on a single leaf. One small fern leaf may produce three or four million spores; a large plant may form twenty or thirty million spores—each a potential new plant. So far from fern "seed" being rare, it is a characteristic example of the prodigality of plant life. If every spore of an ordinary large fern lived and came to maturity, there would be, let us say, 25,000,000 new fern plants.

If we suppose that each spore had landed far enough from

its neighbors so that all these plants might live and each new plant occupied about 1 square foot of land, this population, the progeny of a single parent, would cover about 600 acres. If the whole process were repeated once more, every one of these ferns again producing 25,000,000 new ferns, the resulting greenery *would cover the entire continents of Europe and Asia*! If we compare this theoretical world of ferns with their actual distribution over the earth, it is evident that the mortality of fern seed runs into incalculable billions.

As recently in the history of botany as the middle of the nineteenth century, at a time when the vegetation of Europe was well catalogued and Asa Gray had written his *Manual of the Botany of the Northern United States*, botanists were

much exercised over the origin of fern seed. The spore-sacs they regarded as comparable to the pistil of a flower; but where were the stamens? Reasoning by analogy, they were convinced that the "seed" had its origin, like that of flowering plants, in a union of cells; but what could they liken to the pollen? Analogy is a treacherous servant. We now know that they were quite wrong in their assumptions and were wasting their time when they attempted to see stamens in various glandular hairs and appendages on certain fern fronds.

Fern spores originate in a way which is more comparable to the origin of the pollen itself; in their history is no union of cells. Gather a pinch of spores on the point of a penknife or scalpel, drop them into a dish of water or into a nutrient solution made by dissolving certain salts in water. In a few days the water will be covered with a film of green. The microscope shows that the green scum is composed of thousands of minute plants, one from each spore.

In a few more weeks you can recognize the individual plants with the unaided eye; each floats on the surface of the liquid, a small green disc with edges more or less lobed and ruffled. We can follow the story of development through the microscope. The rough hard wall of the spore first cracks open, the living protoplasm emerges. The protruding portion becomes divided into a terminal and a basal cell, the former containing an abundance of chlorophyll, the latter being usually colourless or nearly so. The terminal cell divides again and the cells thus formed repeat the process, until a plate of green cells has been formed. From various cells, usually those nearest the spore wall, slender transparent hairs sprout, as it were the simplest possible type of root. As enlargement continues, one group of cells at the margin is set aside as a meristem, a region of active and continuing division; around it cells divide, enlarge and mature, surpassing it so that it comes to lie in a notch.

The resulting plant may have the outline of a heart; or other irregularities in growth may develop, so that it becomes lobed, even butterfly-like in form.

And this is all. This plant rarely exceeds a couple of centimeters in diameter; never forms any fronds, any stems or proper roots. It may grow slowly for many months, inhabiting perhaps the cool moist rocks next to a running brook. I have seen them by the hundreds in the damp clay on the floor of a large cave, all standing up on their rear lobes and facing the source of the dim light. They have no means of tapping hidden supplies of water and dry air is quickly fatal; they cannot compete with hardier plants in breezy and well-lighted places.

But are these ferns? Is this feeble plantlet all the outcome of the complex reproductive structures of a fern frond? What is to become of a race of plants whose offspring resembles them no more than this does its parent?

This delicate creature forms very characteristic reproductive structures which are unlike anything produced on the fronds and, moreover, bear no obvious resemblance to anything formed by a flowering plant. And the plant which it forms through the activity of these structures is obviously a fern, a fern that we recognize as such, resembling, if not its own parent, the parent of that parent. Thus is the cycle completed, the species maintained.

On the underside of the flat little plant, projecting into the thin film of water that lies between it and the rock or soil to which it clings, are its means of propagation. There are small dome-shaped bodies, whose tops are detached as lids and from whose interior swarms of minute creatures emerge into the water. These microbes swim away actively in all directions, each tumbling over and over as it moves. If it is stopped in its career, it may be seen—when sufficiently magnified—as a corkscrew, a spiral of transparent protoplasm. Its motive apparatus is a tuft of lashing hairs —flagella, whips—attached near one end.

Fig. A Corkscrew, a Spiral of Transparent Protoplasm

Elsewhere are transparent projecting chimneys, mostly slightly askew. Down the centre of each of these is a narrow canal, into which the swimming germs enter if perchance they come near. At the bottom of the canal they find a large immobile cell and with this cell one of them unites. Here is the analogue of the fusion of gametes which takes place in the ovule of a flower and begins the life of an embryo:—here next the ground on the lower side of a small flat plant and not on the frond where the older botanists looked for it.

The structures here involved are of more than passing interest to the botanist. The dome in which the swimming cells are generated is named antheridium; similar structures are found in many flowerless plants. The structure which contains the immobile cell and in which the union of cells occurs that forms the embryo, is an archegonium; archegonia are found in a wide variety of plants which seem to have little else in common—plants so various as mosses and pine trees. The botanist, contemplating this unexpected similarity among plants which are not obviously related, is apt to assign to the structure in question considerable significance. Such an identity of structure, he reasons, cannot have arisen in so many dissimilar plants purely by accident, by coincidence. It is easier to suppose that these plants are really related, perhaps very far back in their ancestry and that their differences have developed since then.

On such a basis are founded such groups as the Archegoniatae, in which were placed all plants with archegonia: an immense and varied assemblage. We now know that the development of species does indeed yield the most surprising resemblances by a sort of coincidence which is called convergence; that such resemblances are not wholly to be relied upon.

We therefore take with a large grain of salt conclusions about relatedness which are based on single characters. It is probable, however, that most of the varied "Archegoniatae" really did spring from the same extremely ancient stock, in which the germs of archegonia and the resulting method of cell-union had taken shape.

Fig. A New Fern Plant

The new fern plant, the embryo, has its beginning in the cavity of the archegonium. Slowly, through weeks and months, it sends down a tiny root, sends up a miniature leaf. These break out of the parental tissues, the leaf curls around the edge and up to the light. Then more roots, more leaves. Finally a stem, from the tip of which larger leaves, recognizable fern fronds appear. In moist and mossy clefts of rock, under overhanging ledges in deep wooded ravines, on sheltered patches of mud in caves and thickets, these small tufts of leaves may be found, unfern-like in appearance and often puzzling to the amateur botanist.

Their growth to mature fernhood may be a matter of many months. As they develop, their parents, the flat small plants to which they are still attached, gradually wither and die; the support of such a monstrous progeny is too much for them and they are done to death by their own offspring. One of these plantlets bears usually only a single embryo. If, however, this embryo is removed, another will often appear, growing from some other archegonium where its elder brother kept it dormant. If all embryos are kept picked off as fast as they become visible, the parent plant, the flat plant, may reach the —for it—monstrous diameter of several inches.

If now the entire sequence of fern reproduction is reviewed, a noteworthy statement must be made: *Parent and offspring are never alike,* in apparent contradiction to all one has come to believe true of the lives and generations of plants and animals. They are so different that they would never be recognized for members of the same group of plants, far less for members of a single species.

One is large and leafy, possesses complex conductive tissues and reproduces by forming spores without union of cells. The other is small and flat, lacks internally differentiated tissues and reproduces through the union of cells in special containers which resemble nothing found on or in its parent. The result of the reproduction of either type of fern-plant is the appearance of the other type of body. A fern species exists in two guises, more different than the fictional Jekyll and Hyde.

The two plants differ also in a character visible only with

a microscope: the number of chromosomes in each cell. The larger plant, the leafy fern, is diploid like a flowering plant, it has two chromosome-sets per cell. The smaller plant is haploid, like pollen. The reduction is accomplished when the spores are formed, on the leaves and in the spore-cases. The increase occurs when the gametes unite.

It is interesting that parent and offspring, which differ so strikingly in appearance, differ in the quantity (not necessarily the quality) of genes in their cells. Interesting but apparently not decisive; for sometimes the large gamete in the archegonium grows into a new plant without fertilization. The new plant then has the same chromosome number as its parent, but differs in the usual way. Such phenomena are known also amongst flowering plants. It is possible that the change in chromosome number is concomitant with some deeper change which is responsible for the visible differentiation.

One of the most intriguing aspects of the entire cycle of events in the reproduction of a fern is that it is not very successful. The smaller of the two generations of a fern is a weakling. It is easily injured by strong light and killed by even short exposure to dry air.

Furthermore it clings to flat places instead of rising erect as most of its competitors do; it is easily smothered by more aggressive types of vegetation. Not only is it intrinsically ill adapted to survival in all but a few types of environment, its reproduction depends upon a chance supply of moisture from the outside, in which the swimming germs may travel to the egg cell; in flowering plants, in contrast, this journey is accomplished in the internal and never-failing liquid of the pollen tube.

Even when fertilization has been accomplished, the growth of the embryo is slow and for many weeks, even months, the young plant must make its way without any protection against the adversities which are so easily fatal to young plants. Compare the embryo of a plum seed, surrounded by seed coat and by the stone and flesh and skin of the fruit; or that of a grain of wheat, packed next to a rich supply of food sufficient for several weeks growth and almost all embryos in seeds have already formed their first leaves, stem and root-tip before they even emerge into the dangers of the world outside.

It is for such reasons that the flowering plants occupy most of the surface of the earth and the ferns are limited in range. The history of the earth shows us that once upon a time ferns were extremely abundant, indeed the dominant group on the earth. Their abundant remains, metamorphosed into coal, still furnish us energy. The flowering plants were later arrivals on the scene; but, when they did appear, they conquered the world.

Today most ferns are confined to the moist and shady ravines and the swampy forests of the north and to the tropical forests of equatorial lands. In certain tropical forests, indeed, especially on islands swept by mild and moisture-laden clouds from the oceans, ferns attain stature and abundance comparable to their former prosperity; some of the tree-ferns of New Zealand and the Hawaiian Islands suggest the fantastic scenes conjured up by the paleobotanist.

In such plant paradises also ferns may be found perched in the crotches and on the limbs of the great trees, forming masses of tangled roots which catch the abundant water which cascades down the bark. They (with orchids, bromeliads and others which share their choice of situation) are not parasites, though often referred to as such; they take no nutriment from the tree to which they are attached.

In the moist and cool climates of northern Europe and northwestern America also ferns can flourish; the big bracken ferns often cover hills and mountain-sides with their sweet-smelling leafage, a grateful cover for rabbits, deer, boys and other small animals. The hay-scented ferns invade pastures in New England and maritime Canada. A few ferns can cling to faces of rocks and there endure exposure to sun, wind and ice; they are small and usually leathery or woolly plants with wiry stems.

Ferns are favourite victims of the human passion for domestication. Their "wild and fragrant innocence" must be confined within doors for the adornment of our dwellings.

That they do not flourish in captivity is to be ascribed usually to the atmosphere with which humanity surrounds itself. The habitants of the cool moist forest do not take kindly to the parched hot air of a steam-heated apartment. This is usually the reason that the leaves die back at their tips:—the common superstition that to touch a fern leaf is to kill it is unfounded.

Fig. Fern Fronds (Osmunda)

Where ferns do grow, they usually grow luxuriantly; a forest floor carpeted with great Osmundas, or a ravine filled with shield-ferns, is a green and pleasant place. Most of this abundant vegetation is leaves: leaves so large that botanists have hesitated to call them leaves and have called them fronds instead—another word which means the same thing.

An ordinary fern is mostly leaves, leaves which are often lobed or divided into many small leaflets, these in turn often again lobed or divided. Such compound leaves may simulate erect stems which bear many small leaves and indeed, in some respects they are more like a system of branches than they are like the small leaves of apple-trees or rose-bushes.

Fig. Walking Fern (Ravine)

Common ferns of temperate lands have a stem no bigger than your thumb, a small knob covered by the bases of the leaves and often clothed with hairs or scales. From the lower part of this stem spring small wiry roots. In spite of it diminutive size this stem has within it the destiny of the whole plant.

At its apex is a region of growth, from which springs every season the noble crown of fronds which unfold in the spring and replace the dying or dead leaves of the year that is past. Everyone knows the uncoiling fronds of April and May and the country-bred hunt them as edible delicacies. The method of unrolling is characteristic of ferns and of few other plants.

In the stem also is the food which enables the plant to live through its leafless seasons and to send up its periodic cluster of new foliage. Ferns depend on their stems as flowering plants do on fruits and seeds, to get them through periods of dormancy. Some ferns have more lengthy stems which grow just beneath the surface and parallel to it, often forking and forming two crowns of leaves. As it grows at one end such a stem dies off gradually at the other and thus creeps along through the soil from year to year. In the tropics the stems of many ferns stand upright, may reach heights of thirty or forty feet.

At their tips spread the great leaves, which may be six or eight feet long, or more. Immense as is the surface of such leaves, their internal anatomy is quite simple, the cells in few layers and ill protected against rapid evaporation. It is easy to understand why such delicate films of chlorophyllous tissue could not long survive the storms and droughts of "temperate" climates.

Fig. The Walking Fern (Camptosorus)

One of the most individual of ferns is the walking fern (Camptosorus). It has an undivided leaf, a gradually tapering structure something like a spearhead. The points of some of these leaves continue to grow to an amazing extent, running out into long green hairs; at the extreme tip is a growing region, in which new parts originate and are added to those behind. (It is this feature, found to some extent in all fern leaves, that recalls the growth of stems from their terminal buds.) Finally the growing point of Camptosorus apparently finds itself too far away from the rest of the plant to which it belongs and connected by too tenuous a thread.

It organizes itself into a new plantlet, an embryo, which consists of new miniature leaves, a stem, several minute root-tips. If a little soil or wet moss is haply encountered by the drooping tip, this small plant, literally an offspring, takes root and sets out for itself. Eventually, perhaps while it is still attached to its parent by the long and withered leaf-tip, it forms tapering leaves of its own which repeat the process. A succession of arching fronds may thus "walk" over a hillside or scramble up a rocky bank. Ultimately the connecting threads disappear completely and reproduction has been accomplished, without benefit of spores, uniting cells, or any special machinery.

Another fern (Cystopteris) sprouts small outgrowths here and there on the leaves; usually on the under side near the veins. Each is made of two thick short leaves with a growing point between; a bud, in fact, or a bulblet. They are easily detached; they fall to the ground, or perhaps into a rivulet which carries them to fresh pastures. Each can resume growth, form new and more fern-like leaves and come finally to resemble its parent.

Many other ferns manage to proliferate in one or the other of these ways, or in ways similar to them. Ferns are commonly propagated simply by breaking apart the branches of the short underground stems; a process which occurs in nature when the death that creeps up on the older regions of the stem reaches a point where a forking occurred and separates the two growing points into two individual plants by causing the disintegration of the tissues that have held them together.

Indeed ferns are so apt at multiplying themselves and even at migrating without recourse to the complicated sequence of spores, archegonia and antheridia, that in this way they make up for the poor success of their normal reproductive structures. No complex special parts are brought into play, no visitors from the animal world; no special environment is requisite.

That many-sided capacity of the plant for growth at apices and at other points manifests itself in new outgrowths, in parts which can later become detached and survive as separate individuals. Ferns in this respect are representative of the plant world; animals are much less versatile in their growth and reproduction. Starfish can replace amputated rays, salamanders can grow new legs if necessary. But even a giraffe does not continue indefinitely its apical extension and its head, so distant from its body, does not sprout new legs and digestive system of its own and part company from the old.

"Mosse," wrote Francis Bacon, "is but the rudiment of a plant and (as it were) the mould of earth, or barke. The mosse of trees is a kinde of haire; for it is the juyce of the tree, that is excerned." The popular idea of moss is still not far from the Lord Chancellor's. If plants are small enough, without easily visible leaves or stem, a green woolly mat on the earth or on fallen logs or on the bark of trees, or a green scum on water, or small plants which grow in aquaria for the benefit of goldfish, they are spoken of as moss and treated with scant respect and perhaps some doubt whether they are really alive or not.

Different kinds of moss have received popular names: reindeer moss grows in the far northern tundras (and elsewhere), Spanish moss hangs from the boughs of trees in southern states, Irish moss is made into puddings along the shores of the north Atlantic, clubmoss is sought in the woods for Christmas decorations.

Careful as we may be not to run rashly counter to established popular usage, here is one concept so vague as to be of little value to any one. A word that means any small plant or any plant that you do not care to look carefully at is a word that is next to useless. The distinctions of the botanist are here neither pedantic nor frivolous and by following his usage we may clarify and enlarge our vision. Clubmoss, also called ground pine, is related neither to pines nor to anything else that is called moss, certainly not to moss roses (which are not roses), Spanish moss (which is a flowering plant), reindeer moss (which is a lichen), or Irish moss (which is a seaweed). Pondscums are commonly algae but may include those small flowering plants spoken of as duckweed.

Aquarium moss usually belongs to the flowering group. But besides all these and related to none of them, is a large class of small plants which, though common in most parts of the world, are scarcely regarded as individuals or as alive and are referred to contemptuously as moss, without qualification. These are the mosses which make tree-trunks and walls green, which form green cushions on stony ledges and clothe with green the rocky beds of mountain streams, which cover fallen trees and the debris of the forest floor in a green blanket. These are the plants which the botanist chooses to call the true mosses (Musci).

Small as these plants are, they have parts, which are clearly revealed by an ordinary hand magnifier. Their diminutive stems may stand erect or trail over the substratum. They bear close-set small leaves which may be flat, oval or lance-shaped in outline, or rigid and needle-like; usually less than a centimeter long. The plants may grow so tightly packed as to form a cushion which resembles green plush, or a treacherous spongy mass which may float on deep and still water; or they may stand in groves like minute pines.

When we refer to their parts as stems and leaves we may cheat our understanding; they are not like the stems and leaves of flowering plants or ferns,—they have no highly specialized conductive tissues within them, no epidermis with stomata leading to a system of internal air spaces. Most of the leaves are simply plates of cells, one cell thick and nearly uniform in structure. Some have a central rib and the most elaborate have vertical plates of cells rising from the upper side of the midrib and extending the length of the leaf.

The stems usually have a strand of somewhat elongated cells in the centre and perhaps around this thick walled cells which render it tough and strong; there is nothing corresponding to the elaborate water-conducting vessels of a geranium stem, no complex pattern of many differentiated tissues. Water rises easily the short distance to the top of a moss plant without special conductive pipes; in fact, it has been shown that most of their supply rises outside them, on their surfaces, in the chinks between neighboring plants and tightly pressed leaves.

Resistance to stresses and strains is not an urgent need for plants more than an inch high. As far as food is concerned, they lay up no store for the morrow, form no tubers or fruits, but live each day on the sunlight thereof, becoming readily dormant during unkind seasons.

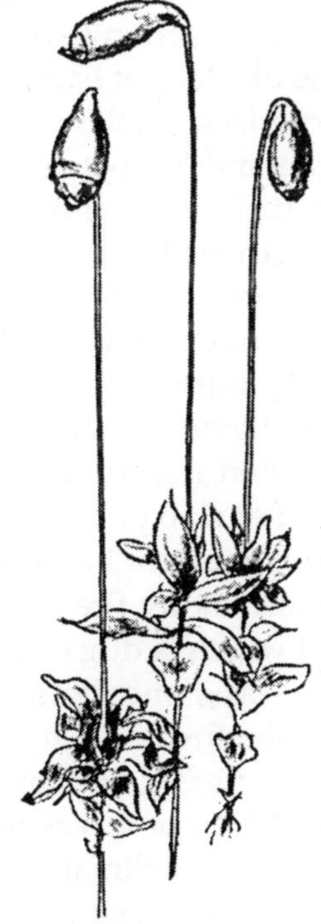

Fig. Mossess

Perhaps the most remarkable thing about mosses, especially in view of their extreme structural simplicity, is their capacity to endure punishment. In the mountains they cover rocks and trees, embrace the cascading water, in regions of a pleasant and moist coolness—while summer lasts. But summer lasts only three or four months.

The rest of the year the mosses spend swathed in ice, blanketed in deep snow. Frigor seems not to trouble them. All those frozen months they live on in a state of suspended animation and resume activity when warmth returns. Elsewhere, as along our western shores, they flourish through a mild and moist winter and spring and dry up at the coming of the and summer. Crisped and brittle, they cling to rocks and to the bark of trees through the warm dry months and come to life again when the rains of autumn soak them.

Some mosses live completely submersed, many feet beneath the surface of lakes, forming soft beds a foot or more thick. Others inhabit the bleakest of bare rock walls, little dark patches of dirty green or black wool. Any one species is rather limited in its choice of a home; but between them all (and there are many species) they can cover with their delicate verdure almost any spot on earth wet for some part of the year, providing that larger plants do not evict them.

In a house on Dartmoor I have seen a fire smoldering which they said had been burning continuously for at least a hundred years. It never flamed high, but glowed redly and gave off a good heat. When it burned less brightly, those of the house threw on another slab of peat. Everywhere in those northern countries peat is used for hearths; black fibrous stuff with a characteristic odor, taking fire and smoldering slowly and hotly; valuable fuel which is dug out of the ground, often not far from the surface;—as if one should dig the sod of one's garden and throw it on the fire.

Peat, like coal, is the remains of plants which grew many thousands of years ago. Much of it was formed from mosses, which grow well in cool damp climates and readily form, in very wet places, thick beds which soak up and hold the water like sponges. Such mossy pastures still wet the traveler's feet on the high moors of Wales and Devon and the dreadful muskeg of the northern wilderness has a similar origin.

Under warmer and brighter skies the older portions of

the mosses would die and decompose, returning into the dust and air of which they were constructed. But where they are cold and continually soaked with mist and rain, they die but do not disintegrate; they form peat. This half-decayed organic material becomes compressed by the weight of living plants above. If the climate changes, if it becomes drier, soil may be formed on the surface. Finally it may be found and dug out by Celt or Gael or Teuton and used to replenish the eternal fire on his hearth.

Some of the peat-beds are very old. They contain remains of many other plants besides mosses, often spores and pollen and other small fragments which can be identified as closely similar to those of known species which live today. Scientists have amused themselves by sifting small pieces of peat and identifying the kinds of plants that grew in that part of the world before the great ice-sheets came down and covered the land. In this way the aspect of vanished ages may be reconstructed.

The capacity of mosses for reproduction equals their ability to survive the turns of the seasons. There are no flowers on them. But all sorts of small buds or hair-like protrusions may appear from almost any of their parts and grow into new mosses which may become detached as separate individuals. It is by such methods, indeed, that the great cushions and beds of moss are formed. Pull off a leaf of moss and place it in water. The chances are that in a few days it will sprout green hairs, filaments of cylindrical cells arranged end to end.

On these, at their tips, buds appear, small oval masses of cells which grow into new leafy branches and which by the disappearance of the hairs that bore them become distinct plants. It has been said that if you put a mass of moss through a grinder of some kind (like those used for grinding coffee or meat) every uninjured cell can grow into a new plant. Some mosses make little disc-like objects at the ends of their stems in a little cup formed of the surrounding leaves; these are easily dislodged and washed away by splashing raindrops and wherever they fall can grow, becoming complete new plants.

At the tips of the stems of certain mosses, at certain times of the year, one can find antheridia and archegonia comparable to those of a fern; the moss plant corresponds in this respect to the smaller of the types of plants that alternate in the life cycle of a fern. The archegonia are longer and slimmer than those of ferns; they are shaped rather like the glass chimney of an old-fashioned kerosene lamp, the egg cell being contained in the swollen part just above the base. Many archegonia stand crowded on the tip of a stem, like the hypothetical angels on the problematic head of a pin, surrounded by closely clustering leaves and frequently immersed in water which has risen among the leaves or fallen from above and is held in that tiny goblet.

Fig. Moss Flowers

The antheridia, somewhat cigar-shaped, stand similarly at the tip of the stem, sometimes mixed with the archegonia but more often on separate branches or even on separate plants. The leaves at the apex of such stems commonly spread apart to form a rosette, the so-called moss flower. When the antheridia are covered by water (which may be held for some time by the swollen hairs which grow among them), they break open and discharge multitudes of their swimming germs, essentially like those of ferns.

Drops of water which swarm with the tiny swimming creatures may be splashed or otherwise moved to the vicinity of a cluster of archegonia, where a few of all those innumerable germs of life can fulfill their destiny, pass down the canals of archegonia and unite with the egg cells within. A new plant is created in the swollen part of an archegonium.

Observe what happens to the new moss, the embryo thus generated by the union of cells. It begins life in the archegonium, still atop the parent plant. It grows at once into a small rod and as it becomes longer the surrounding part of the archegonium also grows longer and continues to envelop it. This continues until the basal part of the archegonium may be three or four times as long as before. At last the pace becomes too rapid.

The neck of the archegonium is broken off and carried up on the tip of the furiously growing embryo within, which now becomes visible as a slender green hair. The lower part of the archegonium remains in place, a socket for the swelling lower end of the still growing embryo. The new moss plant is now a green stalk, its foot embedded in the tip of its parent, wearing on its head a conical hat likewise derived from tissues of its parent. Within this hood its tip swells into an ovoid or cylindrical capsule. This commonly turns to one side, droops; the parental hood usually falling off.

And this is all. This almost microscopic object, perched upon the parent stem and in effect little more than a continuation of it, is the result of the reproduction of a moss or of two mosses, of the conjugation of their reproductive cells in the archegonium. Here is the stage in the life cycle which corresponds to a fern plant as we ordinarily know it.

The capsule and its stalk had their beginning in the union of cells, in a zygote, just as the stem and the leaves of a fern plant had. And in the capsule spores are formed, spores which again will yield the type of moss with which we began this history. So it is possible to regard the stalked capsule as a separate generation of the moss, the spore-bearing generation, in spite of the fact that it derives much of its water and even some of its nutriment from its parent and never leads what one might speak of as a life of its own.

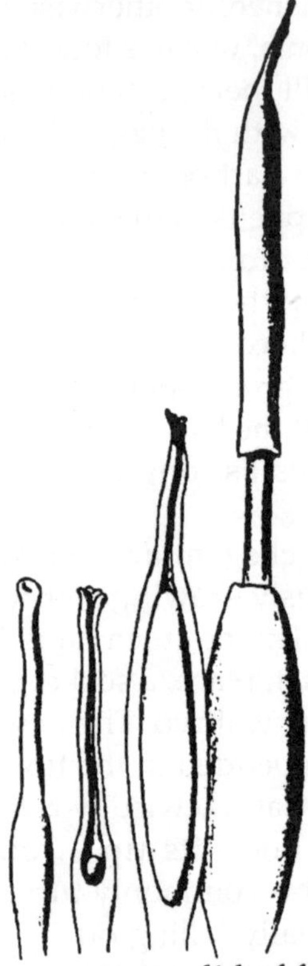

We can say of mosses as we did of ferns that they exist in two guises, one leafy, the other lacking even minute leaves and consisting simply of a reproductive sac on a pedicel. These two differ also in chromosome number as the successive generations of a fern differ:-the gamete-forming generation being haploid, the sporeforming body diploid. Perhaps such a comparison is a bit far-fetched, a bit artificial; but on such considerations elaborate theories of the descent of plants have been based.

The capsule is an intricate and beautiful body. After that conical covering derived from its parent archegonium has fallen, a small lid becomes visible, which in turn becomes detached. Now a ring of teeth is seen; often a second ring within. These teeth are among the marvelously beautiful things of the plant world—if one has but a microscope to see them withal. If you breathe on them they fold together; as they dry they curl outward like the points of a crown. By their number and arrangement, by their minute markings, mosses are classified and recognized. Fascinated students have been led by the beauty of the peristome (thus are the teeth designated) into devoting their lives to the study of the Musci.

Within the capsule, among other things, is a sac which contains spores; thousands of spores, much like those of ferns. These drift out between the teeth when the air is dry and the teeth spread apart. They float around on the breezes and gradually settle to the ground; millions of these potential mosses are scattered extravagantly over the earth every year. Now and then a few may chance to settle in a patch of water or moist soil where they can grow. They become minute branching green threads and on these appear buds which become stems and leaves. The cycle is joined again.

A group of plants related to mosses and not quite so universally common is called the liverworts (Hepaticae). Many of them are like the mosses in general appearance, speak of as a life of its own. We can say of mosses as we did of ferns that they exist in two guises, one leafy, the other lacking even minute leaves and consisting simply of a reproductive sac on a pedicel. These two differ also in chromosome number as the successive generations of a fern differ:-the gamete-forming generation being haploid, the sporeforming body diploid. Perhaps such a comparison is a bit far-fetched, a bit artificial; but on such considerations elaborate theories of the descent of plants have been based.

The capsule is an intricate and beautiful body. After that conical covering derived from its parent archegonium has fallen, a small lid becomes visible, which in turn becomes detached.

Now a ring of teeth is seen; often a second ring within. These teeth are among the marvelously beautiful things of the plant world—if one has but a microscope to see them withal. If you breathe on them they fold together; as they dry they curl outward like the points of a crown. By their number and arrangement, by their minute markings, mosses are classified and recognized. Fascinated students have been led by the beauty of the peristome (thus are the teeth designated) into devoting their lives to the study of the Musci.

Within the capsule, among other things, is a sac which contains spores; thousands of spores, much like those of ferns. These drift out between the teeth when the air is dry and the teeth spread apart. They float around on the breezes and gradually settle to the ground; millions of these potential mosses are scattered extravagantly over the earth every year. Now and then a few may chance to settle in a patch of water or moist soil where they can grow. They become minute branching green threads and on these appear buds which become stems and leaves. The cycle is joined again.

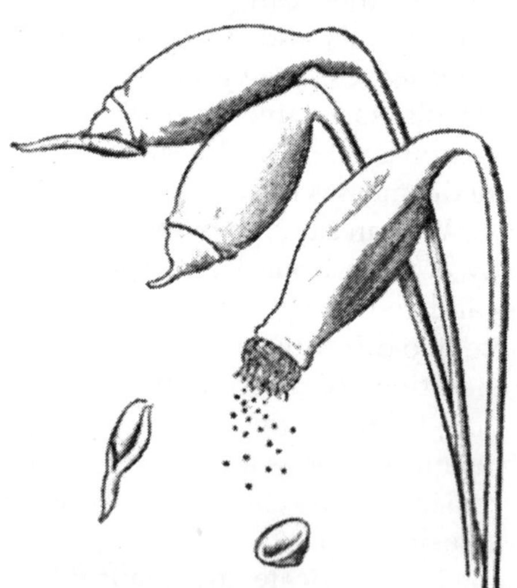

Fig. Moss - Capsules

A group of plants related to mosses and not quite so universally common is called the liverworts (Hepaticae). Many of them are like the mosses in general appearance, small trailing stems covered with minute leaves.

The leaves are simpler even than those of most mosses internally, but those of many species are elaborately toothed or divided and may possess curious sacs in which water collects and which are sometimes inhabited by minute aquatic animals: a balanced aquarium in a leaf! Some of the liverworts have evolved into rather curious types of plants which are quite unlike the mosses in appearance: flat plates or forking ribbons of green, usually growing prostrate and overlapping on rocky ledges or patches of mud, in just those situations likely to be inhabited by the smaller of the generations of a fern.

Within that flat and apparently simple structure is sometimes a marvelous house of many mansions, green-walled, roofed with translucent green cells, their floors often covered with crowded green hairs. Each chamber opens to the exterior world above it by a hole, often curiously fashioned like a small barrel without top or bottom. The plants creep forward over the substratum by growing at one end and dying at the other; as they grow forward they fork, form rosettes of slowly migrating sister plants. From their lower surfaces thin colourless hairs reach down into the chinks of rock or soil.

These plants are considered to be related to mosses and to the more moss-like liverworts because of the close similarity of their archegonia and antheridia and of the small capsules which represent the spore-bearing generation. In some of the simpler liverworts archegonia and antheridia are formed anywhere on the surface. In the more complex kinds they occur on special branches.

Fig. Liver Wort

In one common kind (*Conocephalum*) the antheridia are

embedded in round flat discs, the archegonia occupy the under surface of a small knob with a conical head; a unique and curious structure in the plant world. After the union of cells and the origin of the embryo, the stalk of this branch elongates enormously and rapidly, so that the cone-shaped head shoots up into the air.

The small spore-bearing capsules hang from the lower surface of the cone. In still another familiar liverwort, *Marchantia* (familiar, that is, to the naturalist who knows where to look for small plants), both kinds of organs, antheridia and archegonia, grow on stalked discs. Those which bear archegonia are decorated with curious curving green rays arranged like the ribs of an umbrella; after the growth of the embryos in the archegonia these straighten and spread apart like the spokes of a wheel.

Nothing may seem less alluring as a subject for close study than the green scum on the surface of a quiet pond: ooze made of thousands of minute slimy threads tangled together, foamy with imprisoned bubbles of gas. Nothing is more surprisingly beautiful than these threads when they are seen through a microscope. Each thread is composed of living protoplasm, which is enclosed by a cylindrical wall, transparent as a glass tube.

The brilliant green colour may be contained in biscuit-shaped bodies which line the wall and obscure the interior, or in an irregularly branched network similarly pressed against the inner surface of the curving wall. In many of these plants the chlorophyll-containing bodies are of elaborate shapes not seen elsewhere in the plant kingdom: flat girdles extending around three-fourths of the circumference of the thread; paired stars hanging in the midst of the interior space; beautifully frilled ribbons coiled spirally within the wall.

Some of these thread-like living plants are formed of cylindrical cells joined end to end. The centre of each cell is commonly filled with water, in which a small mass of protoplasm may hang suspended by delicate strands which run out to the protoplasm next the wall. In this centrally hung protoplasmic island is the nucleus; like a spider supervising her business from the centre of her web. Other kinds of filaments lack any cross walls to divide them into cells; the thread is a continuous tube filled from end to end with living protoplasm, in which drift multitudes of nuclei. Such a multinucleate protoplasmic body is variously described as a gigantic cell with many nuclei instead of one, or as a large number of cells, each containing one nucleus, not separated by clear boundaries from each other. The first point of view is based upon a concept of a cell as a structural unit, something bounded by walls; the second regards a cell as a physiological unit, a centre of activity. Probably neither of them quite makes sense; the wary botanist speaks of such arrangements as coenocytes.

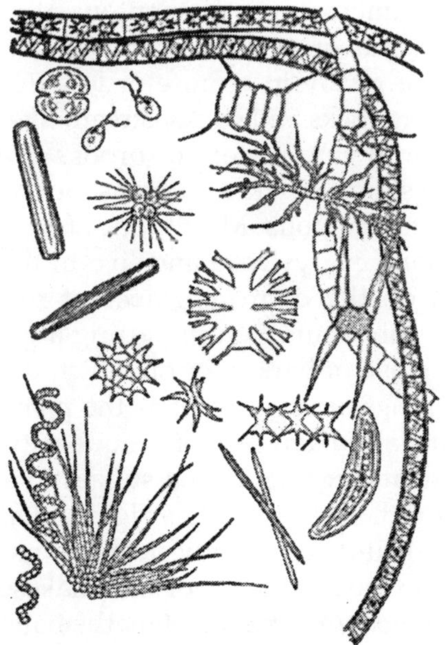

Certain it is that among these small aquatic plants, members of the division of the plant kingdom known as the Algae, we find, not only organisms that do not fit our usual glib cellular terminologies, but some of the most complex cells in the world. Microscopic plants and animals (for there are comparable members of the animal world), perhaps only because they are small, perhaps because they are the furthest removed from dandelions, orchids and man, are spoken of as the "lower" plants and animals. Such an attitude towards the less visible parts of the animate world proceeds partly from evolutionary philosophy, partly from some obscure satisfaction with our own exalted state.

A brief glimpse in a microscope at the varied and almost incredibly ornate forms of some of the minute cells to be picked up in a drop of stagnant water should dispel any notion that the algae and their animal counterparts the protozoa are simple, primitive, or in any way to be sneered at by scientists. *Natura terminos illos, nobile et minus nobile, ignoret,* wrote an acute student of microbiology in the seventeenth century: let nature regard no distinction between high and low.

Some of the microscopic filamentous algae float free at the surfaces of ponds or, with their ends stuck in the mud at the bottom, wave gently in a more or less upright position. Others, attached to rocks and sticks, stream like the green hair of naiads in the flowing waters of brooks or the unresting waves along the shores of lakes and seas. But not all the small green algae are filamentous. Multitudes of still smaller kinds float free in waters everywhere and live in the interstices of moist soil and on the damp surfaces of rocks and trees; individually invisible but often imparting a green tint to the water or to the stone or bark in or on which they swarm. The free-floating throngs are spoken of as the plankton.*

These plants also have cells of unique patterns and often joined in precise arrangements that suggest the concepts of a mathematician rather than the less rigidly defined products of mother earth. Nature constantly surprises us in this matter, showing us—as in the production of snowflakes and of crystals in general—that the fountainhead of mathematics after all may transcend the human brain. It has been said that God is a mathematician; the Algae suggest geometrical leanings.

But not all plants of this group are microscopic. As we use the name—perhaps too loosely—the Algae include also the plants ordinarily known as seaweeds and kelps; the brown, greenish, or red plants, leathery or delicately branched and fringed, which cling to rocky coasts or float on the tides. Here again is a plant kingdom in itself; a vast assemblage of plants, varied, intricate, difficult to study, many of them still imperfectly known to science.

The brown algae make the familiar clothing of the rocks along our northern seas. Their slippery tough bodies hang in thick clusters among the barnacles and limpets, a multitude of small crabs, worms, insects and other swarming life finding shelter and food under their drooping fronds; the high tide covers them, the waves toss them wildly and occasionally tear them from their anchors. Great masses of them drift on the surface, still living and growing.

Some of these brown plants attain enormous sizes. The giant kelps of the Pacific may have a round stalk like a tapering whiplash a hundred feet or more long. The narrow end is at first attached to the bottom by a small cluster of fingers, a "holdfast." The larger en d often is crowned by a thick-walled bladder, a gas-filled buoy which floats at the surface and from which long leaf-like branches extend over the water.

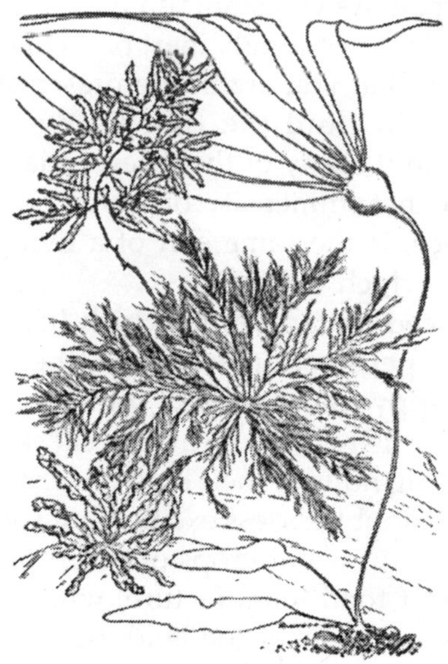

Fig. Sea Weed

Their appearance to the contrary, the plants are "green plants," in the sense that they contain chlorophyll. The brown pigment may be dissolved out and a green frond revealed. Their life is based on the same synthesis of food from inorganic material as occurs in ordinary leaves. Internally the massive plants are rather surprisingly unlike all the other green plants, the flowering plants, conifers, ferns, club-mosses, mosses.

Their leathery fronds, often leaf-like in shape, have no stomata, no palisade layer, no air spaces. Indeed they resemble the stipe in structure, this also being unlike a stem. The entire interior of the body of a brown alga is a mass of interlacing threads embedded in a mucilaginous medium. Near the surface the cells are smaller, contiguous, compose a fairly definite epidermis and a few deeper layers. There is no pattern of bundles of specialized conductive or supporting cells.

The red algae are more delicate, often with a most graceful and feathery appearance. They inhabit mostly the warmer seas and may live at great depths. They also are chlorophyllous, the green masked by the brilliant red pigment.

The waters of the earth teem with the many thousand species of the Algae. As you stand on the shore on a windless day and look over the seemingly inert plain of cold blue water, it is easy to forget that you are in the presence of the mother of life. In that thin fluid the first living creatures were engendered and in it many of their descendants still live and move. Moreover that immensity of water, so inhospitable to our own existence, is the same sort of a living world as this more airy one in which we live.

It is a balanced aquarium, in which life and movement continue at the expense of organic food, the source of which is found in the happy conjunction of sunlight and chlorophyll. Big fish eat smaller fish and these in turn eat smaller ones; until the smallest, finding none of their kind on which to prey, make their living by the doubtless less exciting but somewhat securer consumption of underwater crops of green plants.

The process of food manufacture, which makes possible all the life of the ocean as of the land, occurs under water in conditions quite different from those which obtain in a garden, field, or forest. The light may be quite dim; a few feet beneath the surface it has only a small fraction of the intensity at the surface. Loss of water by evaporation, however, is not a problem. Every cell is nakedly exposed to what light it can get, without fear of complications due to drought. The carbon dioxide is taken already in solution in water, not as a gas from the air. Minerals, inorganic compounds used in the manufacture of protoplasm, of walls, of pigments, are likewise obtained directly from the ambient fluid; there is no problem of transport.

The results of photosynthesis are comparable to those in land plants. Starch is made and from it (or from the sugar into which it may be resolved) the other foods and finally the parts of the plant are constructed. Animals which prey upon the plants and those which absorb these animals, obtain as their reward the organic material which they are powerless to manufacture themselves and without which they can manifest no energy.

Man, the supreme predator, puts his finger into this pie as into all of nature's provisions and removes fish by the thousands of tons. It is unnecessary to take thought for the continuance of fertility in this vast agriculture; the sea is always getting its share of the earth, washed down by the rivers. In fact, the worse man's economy, the more wasteful his culture, the more the sea is replenished. In a field so vast, however, there are naturally local differences.

There may be a temporary dearth of minerals in solution over some hundreds of thousands of square miles and a consequent failure of the crop of Algae. In such regions fish cannot pasture in any abundance and the labour of the fisherman is lost. It is a far cry from the chemical analysis of the nitrogen content of the water to the hunting of whales—but the connection between atoms and leviathans is real and immediate. In this sequence the minute plants of the plankton play the middleman.

Aquatic plants are not only a forest, they are a crop as important to us as those we sow. Millions of tons of carbohydrates are synthesized annually beneath the water—used by man only indirectly, as fish, but none the less used. In certain lands men have learned to replenish the scant fertility of their seaside fields by dragging on to them a harvest of seaweeds; these rot, like the remains of land plants and the products may be used again by new plants.

Moreover, seaweeds have a curious ability to accumulate within their bodies large quantities of the elements which they take from the sea; iodine, for instance. By burning the Algae we can recover such elements in the ash and so the marine harvest assumes importance in the supply of certain of the less abundant elements. Certain Algae are used as food, especially by the fisherfolk of Japan and by the Chinese.

The brown kelps are coarse fare for the poor, the more delicate red Algae are esteemed by the epicure. Their food content is not high, but they probably supply valuable vitamins. We use some also for their medicinal effect. Agar is a staple of the biological laboratory. Much still remains to be learned about practicable methods of using the vast resources of the submersed world.

All these thousands of plants, large and small, visible and unguessed at, reproduce their kind, as corn plants and palm trees reproduce theirs. Their methods of reproduction are as varied, as unique, as complex as their many forms. For an example consider one of the pond-scum algae, a filament of green cells joined end to end. Such a thread of cells grows by processes which compare with the growth of a root or stem.

Transverse partitions divide the cells; the new cells lengthen until each has reached the size of the parent cell; whereupon the division may be repeated. Occasionally, as the filaments push their lengthening coils into an inextricable tangle, they become broken by the waves that toss them up and down and by the fish that swim in the waves and take bites of the Algae; the separated fragments drift apart, each to live its own life and to grow longer by itself. Thus is the species scattered over the waters.

But at certain times and in certain places (the impulse seems to come from a certain conjunction of light, temperature and perhaps other environmental conditions), a cell behaves differently. Its living contents become divided into two protoplasmic masses which lie within the cavity of the parent cell and which form no walls around themselves. They in turn divide into four, these four perhaps into eight and so on until the original cell wall contains a swarm of tiny naked protoplasts, packed together in the space originally occupied by a single cell. As you watch these in the microscope, it becomes evident that each minutia is developing a structure of its own and is in motion; the confined swarm quivers and vibrates.

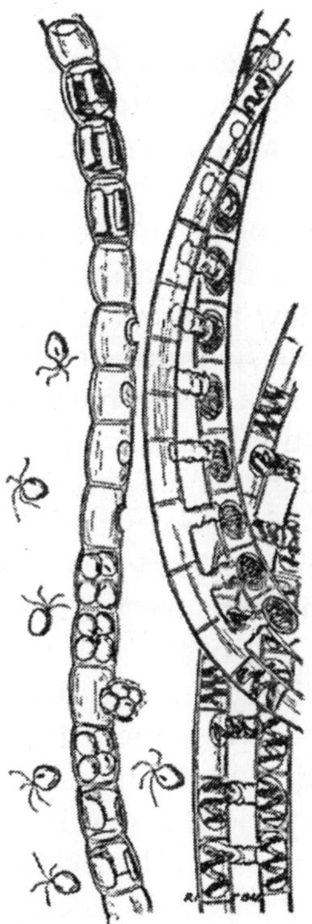

Fig. Reproduction in a Pond

Finally a spot in the parent wall becomes softened; a hole appears and one by one the cells inside are squeezed out. As it emerges, each takes to the water as naturally as a duckling taking its first swim; with minute threads it lashes the water and speeds erratically off into the world. Soon the water around the parent filament is alive with minute swimming creatures, each an individual plantlet on its own, busily rushing around (doubtless as a crowd of men and women at work or play would appear from an impartial planet).

Each is an oval cell without a discernible wall of its own, but containing a green body and a nucleus and, near one end, a curious spot of red pigment which seems to be very sensitive to light. It has been shown that because of this pigment-spot the moving cells tend to swim in the direction from which light comes (unless it is very intense, in which case it affects them in the opposite sense).

Ultimately the small engine begins to run down; motion ceases; the cell settles to the bottom or on a rock or a stick or even on a neighboring algal filament. Being somewhat sticky, it adheres to the surface and there undergoes a remarkable transformation. Its whiplash motile threads are withdrawn, its body lengthens, it is divided into two cells; each of these again divides and a new filament is on its way.

Some of the minute swimming cells fail to develop. Some, of course, form part of a living meal for an aquatic animal pasturing in that multitudinous world. Some perhaps sink to regions of cold and dark where growth becomes impossible. But some are unable to develop by themselves even in the most favorable circumstances. They are unable to develop by themselves; but if they chance to meet a like swarm from another parent, each cell of the one brood may find itself a mate from the other. They unite in pairs; two cells adhering, then the boundary between them fading out, their nuclei approaching and uniting, until one cell only remains. This cell may lie dormant for a period, a spherical cell which surrounds itself with a rather thick wall of cellulose. Ultimately by division it becomes the parent of a new filament or of several new filaments.

The small swimming cells which are able to propagate their kind so readily are commonly regarded as a variety of spore; swarm-spores, they are called. Those that unite and develop only if the union occurs, are obviously gametes, comparable to those of a fern or moss or to those cells which unite in an ovule of a flowering plant. We are in the habit of distinguishing rather sharply between *sexual* reproduction, which is concerned with the union of gametes to make a new germ and those seemingly more facile methods employed by plants (and not so generally by animals), of which the plant grower takes such advantage—the regeneration of a plant from a piece of a branch or of a root or of a leaf, stuck into moist sand and properly cared for.

Here in the Algae, however, we see methods of reproduction which may be classified as sexual or as nonsexual according as they involve or do not involve a conjunction of cells, but which differ only slightly in other ways. Here perhaps is the threshold of sex: a slight differentiation between cells such that each is incomplete in some way that the other can fulfill. What that difference is we do not yet know—though we have some clues. From some such slight physiological peculiarity in certain minute swimming cells the great duality of the animate world has been built up, a division into sexes which differ not only in some qualities of the uniting gametes themselves but in the bodies which form the gametes and in the lives which those bodies lead.

Most of the Algae, indeed, are not so elemental in their sexual life as the species just described. Consider as a further example that beautiful alga which is ornamented with a green ribbon coiled spirally within its wall. Reproduction occurs only when two filaments lie side by side; whereupon small lumps appear at corresponding spots in the walls of the two partners and swell until they meet and form bridges from one filament to the other. The cellulose ends of the joined projections disappear and the bridges are hollow tubes.

The living contents of all the cells of one filament shrink somewhat and flow gently through these narrow connecting tubes into the cells of the other filament, with which they unite. There is an obvious difference between the motile gametes of one filament and the motionless gametes of the other.

This tendency is much more highly developed in most other Algae; which form two classes of gametes as ferns and mosses do:—swimming ones which are discharged into the water and which recall the swarm spores described above and larger motionless gametes retained in some sort of container, into which the swimming gametes make their way and in which the union of gametes occurs. There are no archegonia among the Algae; but there are plenty of bodies which are similar in function.

The differentiation into sexes finds its most complex expression among the large brown and red Algae, which in this as in other features are anything but "low" plants. It is to be noted that when cells unite in the Algae, as in the flowering plants, the effect upon those supremely important cellular constituents, the chromosomes, is always the same. The germ formed by the union, the zygote, is always a diploid cell with two sets of chromosomes.

And somewhere in the life cycle there must occur the converse process, that peculiar division by which the two sets of chromosomes (and genes) are sorted out into single sets in sister cells. In some Algae this occurs just before the formation of the gametes, as it does in the flowering plants; the plant body is diploid. In others the reduction division occurs when the product of the gametes, the cell formed by their union, grows into the plant body; which is consequently haploid.

The workings of heredity are naturally quite different in such species. In still others—and this is an interesting comment on the results of the doubling of chromosome sets—some plants are haploid, some are diploid, all in the same species (as in ferns) and they are alike in appearance; the number of sets of chromosomes seems not to affect their visible characters and they transmit the single or the double number to their offspring according to the type of cell division involved in their growth and reproduction.

Presumably all these differences have arisen in the course of the development of the manifold plant world from its first minute and undifferentiated beginnings. Within this one great group of plants, the Algae, whose very existence is scarcely suspected by the uninitiated, we find illustrations of all the same types of reproductive cycles as we find in the flowering plants and in other more familiar groups of plants—and some besides. The Algae were presumably the first denizens of this world and its only ones for long ages; they foreshadowed all that was to come and attained complexities of existence that have not been surpassed by their larger and land-invading progenies.

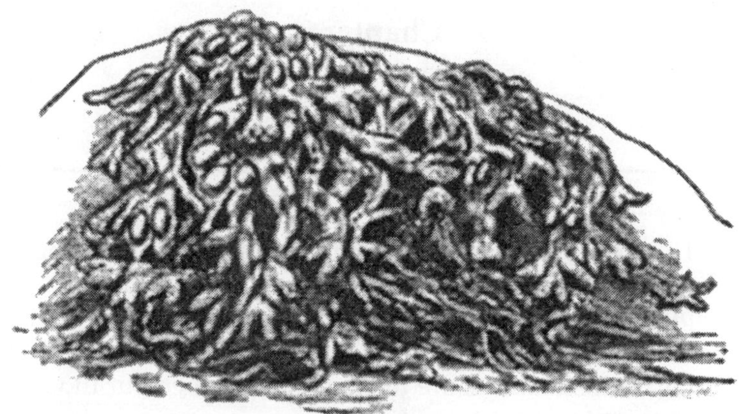

Fig. Rock Weed

Into such curious knowledge are we led by an inquiry into moss and the like,—the common weeds of our forests, mountains and waters. A knowledge not of strange putrefactions of earth which become the rudiments of plants, not of plants arising unsown from soil, nor of plants bearing seed of magic potency; but of exact and complex patterns of life, which diverge from the general formula of the familiar and ubiquitous flowering plants and at present represent a minority faction among plants, but none the less have vast importance in the total economy of the world.

Chapter 9

Botanical Names

To name a thing is the next best thing to understanding it. Indeed, unless you name it, you will *not* understand it. *Nomina si nescis, perit et cognitio rerum,* wrote the father of botanical names: if you know not names, the knowledge also of things is lost. Names are the tools with which our thoughts grasp at nature.

The first man to name plants is said to have been Adam. Though we have no catalogue of the names which he invented, we may suppose that they were the equivalents, in his speech, of names given by our own ancestors: apple, daisy, oak, cowslip. He probably used names much as we ordinarily do, in a rather general way; fully recognizing that for certain purposes it might be desirable to distinguish between various sorts of apple which differed in taste and in virtue. Furthermore, being a horticulturist rather than a botanist, Adam may have been misled by superficial resemblances to make his groups rather too loose and to use one name for apples and other luscious temptations.

We are rich in names of plants, names which bear an imprint of the cradle of the language and of its vigorous youth. The English country names have a quality which no others can share in the minds of Englishmen. "I know a bank where the wild thyme blows," says Oberon, "where oxlips and the nodding violet grows" and the mind's eye wanders over the long grass spangled with buttercups to hedges of thorn in which tall elms stand, to the coppice of holly and oak and to the brick house beyond, timbered with oak and thatched with straw.

Latin names also and Greek, these conjure our imaginations steeped in the vestiges of a Mediterranean culture. Hyacinth and narcissus, olive and fig and plane, these words speak of ox-drawn wains creaking towards the Roman market, of generous vines trained upon elms, of the care of the shepherd for the twin kids dropped by the wayside; of pipes heard in Arcady; of the fauns and facile nymphs who shared those groves.

Latin names in particular have had a curious history. When the tide of Germanic barbarism finally burst the barriers and swept over the luxury and letters of an empire, the art of gardens was almost lost and the knowledge of plants. The newcomers were hunters and robbers and in the flush of conquest their thought was to destroy. But, having the soil and the wealth thereof, they must rediscover the arts that the Romans and Gauls and Iberians and Britons had known.

Here and there survived fragments of Christian Rome, hermits and monks; speaking Latin, preserving the arts of writing, of building, of growing plants for their simple needs. In their gardens they had a few herbs for medicine, a few potherbs for their frugal tables, lilies and roses for the glory of God on His festivals. The Germanic tribes, having names only for common trees and weeds, learned other names from the Latin; corrupted Lactuca into lettuce and achieved parsley out of Petroselinum.

Such ill-bred attempts at naming plants might well suffice for their rude agriculture; a pursuit then, as now, limited to the ignoble or the conquered and rewarded with poverty and contumely. One can grow plants without reading books (though the modern publisher does his best to make gardeners forget this). But there are things about plants not known to the peasant, qualities which must be carefully preserved in the writings of the learned. This is true particularly of their medicinal virtues. Such writing was naturally, in those rough times, done in Latin, the language of the few that were literate.

Since the earliest ages men have known that plants were useful in their ailments and mishaps. In medieval Europe the interest in plants centered about their "virtues." A brew of mugwort (Artemisia) was good for an aching belly and would restore ('twas said) an appetite lost by drinking. Self-heal (Sanicula) was considered to stop bleeding of wounds external and internal. Thistle-stalk boiled in oil was thought to cure asthma or distemper of the ear. And so on through herbal after herbal, an interminable list of traditional and largely mythical remedies.

Such knowledge was in all peoples appropriated by the wise man and the priest, incorporated into religions. Savage forest-dwellers used plant decoctions to make their arrows fatal and surrounded the preparation of the poison with taboo and incantation. The medicine man invoked powerful spirits when he administered his potions. The rabbit's foot or the ceremonial rattle held equal rank with the Cinchona bark, since evidence of the modern clinical kind was sought for neither and both rested their potency on divine intervention.

Fig. Heptica

So also Europeans of five hundred years ago believed that an extract from walnut meats was good for sickness of the brain because of the similarity between the shape and appearance of walnut meat and brain; a beneficent Creator had left His mark upon the plant that man might know the purpose of its existence. Such beliefs survive in our names. Heartsease, which we call also pansy, was known by its heart-shaped leaves; the value of Hepatica might be inferred from its liver-shaped leaves and of all plants the most sovereign was the almost mythical mandrake, whose root had the form of a man and was good for all troubles of the flesh.

It is not surprising, therefore, that most of the earliest books about plants are devoted to their medicinal values. The learned monks and their students made careful manuscript catalogues of plants and enumerated the healing or preventive qualities of each kind. When books began to be printed these compendia were copied in formal treatises.

The science of botany had its roots in the need for descriptions of the plants by which they might be recognized. The early herbals were anonymous, compilations of common beliefs, old wives' tales, sometimes with crude and inaccurate illustrations. Later herbalists began to devote their efforts to careful examination and description of plants; their writings began to have scientific value. These were the botanists from whom our names for plants are descended.

Naturally they wrote in Latin. Their work was often translated into the vulgar tongues for the use of the unlearned and if they occasionally wrote first in the vernacular, their writings appeared also in a Latin version. So have the familiar Latin names come down to us: Rosa, Lilium, Viola, Delphinium, Cornus, Quercus, Vitis and Greek names which the Romans had adopted and spelled in their own way: Hyacinthus, Narcissus, Lotus.

One of the first rewards of persistent inquiry into nature is a distinction between different orders of likeness. All trees are alike, in a general way. All oaks are much more so. And among oaks we can pick out white oaks, red oaks, black oaks, bur oaks and many others. It has been shown that even primitive man made such distinctions.

He arrived at the larger groups either by overlooking small differences (such as variation in the type of teeth on an oak leaf); or by deliberately lumping together things that were perceived as slightly dissimilar. Every child learns, as he grows up, to *name groups,* the size and uniformity of the group varying with the acuity of his perception and with his interest in the objects. The old names of plants, the Latin names and those that have come from the Germanic tongues, are mostly names of what the botanist calls genera: Rosa, Lilium, Acer or Maple, Quercus or Oak,—each of these is a GENUS.

As they attained to greater discrimination, as they began to be interested in the smaller distinctions between plants, the herbalists and early botanists began to name the kinds within each genus; what the botanist now calls a species, what the general public insists on speaking of as a variety. These they at first named by adding a qualifying word to the Latin name of the genus, exactly as we do in English. As we say "*white* rose" or "*red* rose," they wrote "Rosa *alba*" or "Rosa *rubra*" (the adjective naturally following the noun in Latin).

Epithets were sometimes fanciful. Certain trees were distinguished on the basis of a curious notion that some were "male" and others "female" (an idea quite dis-s tinct from the later realization that the parts of flowers are indeed comparable to the male and female organs of animals). So we find two dogwoods named *Cornus mas* and *Cornus foemina.*

Such simple names did not long serve. As new plants came into the catalogues in ever increasing numbers, new names were necessary; this movement was particularly accelerated by the great explorations of the fourteenth and fifteenth centuries. Plantago major and Plantago minor-the greater and the lesser plantains—are names enough for two species. But when John Clayton found a new species in Virginia and Jan Fredrik Gronovius of Leiden (to whom he sent it) tried to name it so that others would know it again, the result was *Plantago foliis lanceolato-ovatis pubescentibus subdenticulatis, spicis cylindricis pubescentibus, scapo angulato;*—plantain with hairy leaves lance- or eggshaped and having very small teeth, the flower-spikes cylindrical, hairy, the flower-stalk angled. This was the name of that species; all of it was necessary, at least for scientific purposes. The botanical works of that period consist mainly of such ponderous brain-twisters. In effect, to name a plant you had to describe it; the more kinds you knew the longer and more complex became your descriptions.

The constant effort at precision had the result which such efforts often have, even in our enlightened time: obscurity. When Leonard Plukenet in 1696 described an American dogwood as *Cornus foemina candidissimis foliis americana* and Philip Miller in 1737 described an American dogwood as *Cornus foemina Virginiana, angustiore folio,* no one could tell whether they had the same or different species; opinions varied.

Since they had not realized how many "female" dogwoods grew in America and how similar they were, they were unable to emphasize the points necessary to distinguish them. The more adjectives they introduced into their names, the more difficult did it become to compare the plants. The result was chaos.

Botany was saved by the great reformers of the eighteenth century. Joseph Pitton de Tournefort was the first to organize into a clear system what we now call the genera; using characters to distinguish them which any one could find and naming each by a *single word or a short phrase.* "Cornus" or "Quercus" was selected as a definite name, whether or not it was accompanied by half a dozen lines of assorted adjectives. Other systematists of that systematic age collated, to the best of their ability, the different descriptive names of the same plant under the particular description which they preferred.

Of these the great Carl Linnaeus was the most successful. His journeys in northern Europe, his industry, his keen observation, above all his powers of classifying, gave him a grasp of the plant world that had not been approached by his predecessors and has not always been equaled by those who have followed him. He wrote several books devoted to the exposition of his system of classification and his rules for naming.

The crowning glory of his literary career was the *Species Plantarum,* published in 1753. In this great work—still the foundation of botanical nomenclature—he recognized and described more or less intelligibly about 6000 species of plants and collected under each the "synonyms," i.e., the names bestowed upon them by his predecessors and colleagues. In this work a device was introduced, possibly only as an afterthought for convenience, which had an immediate and profound effect upon the history of names. In the margin opposite each plant description, Linnaeus wrote a single word (or occasionally two); usually an adjective, perhaps selected from the descriptive name of the species; a sort of epitome of the description, or a reference word which could represent it, used for no other species of that genus.

Instead of having to repeat Rosa *foliis utrinque villosis,* fructu spinoso, to specify that plant you need only say *Rosa villosa*; this would serve to locate the full particulars in the *Species Plantarum* and to distinguish it from all other roses, villose and otherwise. This was the origin of the famed binomial system of nomenclature.

Whether or not Linnaeus himself grasped all the value of these "trivial names," their usefulness was at once perceived by the botanical world. When Miller brought out the eighth edition of his *Gardener's Dictionary* (1768), he inserted the single word *foemina* after the Cornus quoted above, before the long description of the species; henceforth that species was known to gardeners and to botanists simply as *Cornus foemina*. Today every species of plants (and of animals) is known by an equally simple name of two words.

A wonderful system! It is not requisite to invent a single word for each of the hundreds of thousands of species of plants; yet each has a short name which can be readily acquired by those who have business with it. We need only as many single different names as there are genera: Quercus, Rosa, Cornus and the rest are the names of genera—numerous enough, in all conscience, but not impracticably so.

The species can be distinguished by adding to the name of the genus simply qualifying words, which may be used over and over again for different genera: Quercus alba, Cornus alba, Rosa alba, Rosa rugosa, Cornus rugosa, Quercus coccinea, Crataegus coccinea, Ipomoea coccinea. When we run out of descriptive adjectives (as we soon do in a large genus), we can use the names of localities where the species were collected or where they abound: *virginianus, canadensis, missouriensis.* Or we can use a Latinized version of the name of a botanist who collected the first specimen or a friend who helped us to collect it and so perpetuate his memory in a flattering and inexpensive manner: *Cornus drummondi, Phlox douglasi, Monarda bradburyana.* We can also easily name the varieties which we distinguish within a species by simply affixing another word. Such groups are the less stable and less distinct subdivisions of species, particularly dear to the horticulturist: Erica vulgaris is heather, Erica vulgaris alba and Erica vulgaris coccinea are two of the many colour-variants of heather known in cultivation. At first glance this seems like a return to the pre-Linnaean "polynomials"; but a moment's reflection shows that each element of these names is used in a systematic way, one part subordinated to another.

Augustin Pyramus de Candolle, in a textbook published in 1813, wrote as follows: "Thirty thousand species of plants are today known on the surface of the globe; this number would be increased to more than forty thousand, if we added all the undescribed plants in our collections and if we suppose that Asia, Africa and America were known to presentday botanists as Europe was to those of the sixteenth century, we shall see that, in all probability, the terrestrial globe is covered by more than sixty thousand species of plants."

In the second edition of this book, published only six years later, the same sentence appears with the estimated total number changed to "more than one hundred thousand." In the third edition of 1844 it appears that eighty thousand species were known and the estimated total is raised to one hundred and twenty thousand. Botanists now reckon on at least three hundred thousand and admit that many of the most prolific parts of the world await exploration by botanists. Without a concise and practical naming-system, we should long ago have been intellectually buried by such an avalanche of discovery.

So many are the blessings of a concise system of names that we may forget the really greater contribution to science made by Linnaeus. He not only named plants, he classified them. Indeed his system of names was to some extent the offshoot of his classification. He was the first to succeed in arranging all plants in a limited number of groups, these groups in a smaller number of larger groups and so on in hierarchical array.

Species form genera, genera are parts of orders, orders make up classes. Our groups today are not those of Linnaeus and the principles upon which they are founded are not the same as his; but we use a system which has the same general form. Moreover we still use many of the clues which he pointed out in our searching after relationship. We still follow Linnaeus in using the numbers of stamens and of the elements of the pistil as characters of primary importance in classification. Since almost anyone can count, the Linnaean system has been from the time of its introduction a blessing to students intent on placing an unknown flower in its proper category.

The virtue of binomials, apart from their brevity and economy, is that they carry within themselves some information about the grouping and affinities of the species to which they belong. Any binomial beginning with the word Primula—Primula obconica, Primula rosea, Primula veris—at once locates the species in a group of similar species. If every one had a name of its own, quite unlike that of every other species, this handy guide to relationship would be lost.

In spite of the beauty of this system (it has drawbacks, but they are minor and offer no great obstacle to adjustment), there is a popular abhorrence of the scientific names of plants. The greater part of this may perhaps be traced to a fear of foreign or even of foreign-seeming words (this in turn to the current degeneration of spelling and grammar and diminution of vocabulary fostered by certain modern schools). This is a sad reflection upon our language; for Latin words have contributed much to English, much of beauty and of fluency.

Such names as Veronica, Camellia, Delphinium, Rhododendron, Althaea, Narcissus, Daphne have pleasant cadences as well as exact meanings and offer no greater difficulty to a child than some of the more curious etymological survivals of the Gothic tongues. Many indeed are in common use in homes and gardens and florists' shops. It is true that such words as Pteris and Mnium, with their unfamiliar combinations of consonants, presuppose a higher standard of reading than is current in many schools; but the remedy is obvious—and not difficult.

Charles Kingsley wrote: "In law, physic and divinity, folks had sooner be poisoned in Latin, than saved in the mother-tongue." Today he might deplore the opposite extreme. The prejudice against Latin names is characteristic of the intellectually timid or slothful, who promulgate the slander that botany and zoology are little more than vocabularies; that the botanist lavishes an unnecessary erudition upon an unworthy weed and seeks to dull the sweetness of a flower by naming it in an alien tongue. These romantic persons should exercise their eloquence in expounding the aesthetic merits and the valued connotations of such names as horsemint, stinking Willie, pigweed, bunchflower, mudweed, beardtongue.

Many of our wild plants must, for want of an original metaphor, be dedicated to the Father of Evil by those who are afraid of their Latin names; what a pleasure awaits the tourist who finds the devil's paint-brush in the devil's kitchen! Still others of our gracious American wildflowers have borrowed an English name which was used for another species on the other side of the Atlantic. The cowslip which adorns the English springtide bears no resemblance to the Virginia cowslip and as for bluebells—even the Scotch and English bluebells are very different and neither is related to most of the bluebells of America.

It is a pity to conceal the identity of our own flowers with a name that does not fit them, a name that already has its aura of associations; as it was a regrettable nostalgia that fastened upon one of our best-loved birds the name of the English robin redbreast. There are times and places for all things and I am glad that the poet did not write "I know a bank where Thymus serpyllum blows." But the Latin names are not essentially formidable and if they are given as much consideration as the technical vocabularies of baseball and radio need cause no panic.

It must be granted, in any impartial consideration of

scientific names that the scientist has sinned in manhandling language. Many botanists are so unfamiliar with the roots and stems of languages that it is a pity that they should feel qualified to create words. Such names as Kickxia and Sczegleewia do not inspire confidence and *Oscillatoria hahatonkensis* must arouse mirth. But, all in all, scientific names are not much worse than others—and far more useful.

Popular dissatisfaction with botanical names is partly due also to a very general failure to appreciate the system of nature. If while visiting a friend I am invited to look around the garden (and in some places this is inevitable), it may happen also that I am invited to name some of the plants. Not being omniscient, I cannot always name the species before me, but can say only that it belongs to suchand-such a genus or family. The species of some genera which are *ubiquitous* in gardens—Prunus (cherry, plum, peach, almond), Malus (apples), Aster, Crataegus (hawthorns), Acer (maple)—are so complex that they can be accurately known only by the expert in those groups.

Fig. Linnaea

It is sufficient for ordinary knowledge to name the genus and the fact that one can do so should be received with proper acclamation. Even the recognition of the family represents precise and detailed knowledge. Many amateur gardeners and botanists are indeed interested in arriving at such knowledge of their plants and are thwarted by the professional growers, some of whom are interested rather in the more internecine aspects of their art. Horticultural names, consequently, border on chaos.

The principle of hierarchical classification is extremely simple and is in constant use for all sorts of objects. Among vehicles we readily distinguish the horse-drawn and the mechanically propelled; among the latter we can classify some as using gasoline, others fuel oil, still others coal or alcohol. Each of these groups can in turn be broken down.

So we have a number of *species* of roses, Rosa carolina, Rosa setigera, Rosa palustris and the rest, which together make up the *genus* Rosa. This genus and other genera such as Potentilla, Spiraea, Fragaria make up the rose *family*, Rosaceae. This family and the saxifrage family and the bean family and some others compose the rose *order*, the Rosales. On the basis of certain structural features many orders, including the Rosales, are placed in the *class* Dicotyledones. These in turn form a part of the Angiospermae, flowering plants; one of the major *divisions* of the plant world.

There are only a few of these divisions; the one just named is by far the largest. More or less related to it is the division called the Gymnospermae, which includes those plants commonly known as conifers: pines, spruces, hemlocks, firs; the cycads also and a few other remnants of ancient tribes, are usually placed here. A third division (Filices) holds the ferns, ancient and modern. A fourth (Bryophyta) is made of the mosses and their relatives. The fifth is the Algae, those mostly small but sometimes very large and variously coloured habitants of ocean, lakes and streams and the sixth division is composed of plants without chlorophyll, the Fungi.

WORLDS FOR OLD

When the sun rose over the great plains of North America three hundred million years ago, his rays fell upon the countless leaves of a great forest; fell upon them, were absorbed into them, were changed into the energy of chemical transformation. Then as now water from the earth was carried up into the leaves and there joined with a gas from the air to make a food. Then as now animals cropped the herbage for the nutriment that they could not themselves manufacture. The physiological pattern of the world has not essentially changed in all these slow-moving epochs.

But the plants have changed. The browsing cattle of that distant period found no grasses to fill their enormous paunches. Nor were the forests through which they floundered forests of oak and maple and hickory and birch or pine and spruce and hemlock and fir. There were no flowers, no flowering plants; no fruits (in any of the usual senses of that abused word). The largest trees had tall trunks covered with scales arranged in an ascending spiral; they forked and the branches forked, forming a canopy of drooping reptilian twigs. Such plants resembled gigantic club-mosses rather than the trees of our own forests.

Mixed with these were trees suggestive of pines but with narrow flat leaves instead of needles. Still other trees of those forests resembled in everything but size the plants which we call horsetails, or, more learnedly in the equivalent Latin, Equisetum.

Some of these are known as "bulrushes" in various parts of America. We can perhaps imagine how a grove of these plants may appear to a mouse or other small creature; even so the tall trees of Paleozoic Wyoming would have seemed to a botanist miraculously transported thither.

To our eyes a somewhat dismal, even repulsive, landscape, a swampy grove of forking scaly trunks and branches clothed in a scant foliage. The animals that used them as provender suggest the fantasies of a disordered mind; giant reptiles like distended and ill-proportioned lizards, standing erect on enormous haunches and raking in the herbage with short front legs. The graceful furry denizens of our wilderness, the feathered things that swim through our air,—these were as lacking as the bright flowers and gracious fruits which we have come to expect of the plants about us.

The lower stories of that vast forest had an aspect less strange to our eyes; at first glance they seem to have been composed mostly of large ferns. Some of these resembled the modern tree ferns in general appearance: graceful feathery leaves many feet long, crowning low erect stems. Many of them were indeed ferns, though differing in various technicalities from the common run of their modern relatives. Others were seed-ferns: they formed seed on their leaves, seed comparable to the product of a flower. Even a botanist, insisting on his own concept of seed, could have found fern seed in such a forest—and would not have had to wake through the small hours of midsummer night.

Most of these plants were woody, more or less tree-like. They had the same sort of complex internal structure as trees have today: water-conducting xylem, food-carrying phloem, bands and rings of supporting thick-walled cells, thin-walled cells filled with accumulated food; all arranged in definite patterns, different from those typical of modern plants, but none the less complex and doubtless efficient.

We know so much of these plants because we have fragments of them which have endured until this day—turned into stone. The vast plains which they inhabited were swampy. As their broken pieces fell into the mud beneath and were trodden under foot by the giant reptiles, some of them escaped decay and became buried in silt, covered by deposits of rock. Great masses of vegetation became through the vast compulsions of the ages changed into coal, or into oil. Much of it, however, retained its form in stone. These permanent records, autographs, we now turn up by the thousands in arid plains and hills. Some are only impressions, as we see a track made in fresh concrete preserved by the hardening of the cement.

Even though the space thus formed may afterwards be filled with hard material, such a cast can tell us only of the external form of the plant that made it; nothing of its internal structure. Other fossils, however, are real petrifactions. The original plants were subjected to gently running streams of water which, as it washed away their dead tissues cell by cell, replaced them, cell by cell, with rock. Even the parts of the cells may be seen in such fossils. Paleontologists have learned to grind such pieces of rock to extreme thinness; when they are placed on the stage of a microscope, one can study them as if he had a section made from a freshly killed plant. Other methods also have been developed for removing layers of microscopic thinness from the polished surface of a piece of petrified wood or leaf or seed.

There were undoubtedly many kinds of plants of which we know nothing; they left no remains, or they grew in places now submerged and inaccessible beneath the seas. There were presumably herbaceous plants in those days; most of them probably did not resist disintegration long enough to become fossilized. We have traces of smaller plants, something resembling the club-mosses of today; even of a few mosses. Perhaps also some primitive forms of flowering plants were to be seen in that vegetation. The lack of evidence does not disprove their existence. In the main, however, we must classify the plants of 300,000,000 years ago as fern-like, club-moss-like and non-flowering.

In that period, which we designate in our complacent way as primitive and undeveloped (because there were no men there), there were many kinds of plants, in most respects as complex and highly developed as those which in these days make food from water and air. They were different from the plants of today, as different as dinosaurs from jackrabbits; but they were hardly less developed, less differentiated. Most of them have passed away into whatever haven awaits extinct species, without progeny and without trace except for a few beautiful but lifeless fragments of rock, a few veins of coal, a few pools of oil.

It is not yet given us to lift the curtain of the past much further; we can see but dimly into still older periods, regions of misty groping; we have scant and often meaningless fragments, on which we base vague and often careless theories. There were plants before the carboniferous period; of that we may be certain; but we know little about them. We have remains of vascular plants; we have some traces of nonvascular plants something like the modern algae. Perhaps a greater proportion of that vegetation was aquatic and softbodied and left no imprint in the earth's crust.

Perhaps abundant deposits of those older fossils yet await the earnest digger, or lie forever concealed from him beneath the more recent oceans. It is probable, however, that the lack of vascular remains from ancient levels indicates a real difference in those older plant-worlds. It is probable that if we could go back far enough in the history of life, we should find a period, in all likelihood a vast period, during which the only plants, indeed the only living things, were relatively simply organized plants comparable to some of the Algae of today.

At the far end of such dimly envisioned ages it is necessary to imagine some sort of beginning of life. The earth itself had a beginning. Before it was—it was not. Its primeval rocks supported no life. Somewhere, somewhen, somehow, in the turbulent and ceaseless play of atoms some combination was realized which had those peculiar colloidal properties, that character of extensibility, that power of taking to itself water and other things and changing them into its own substance,—that combination of features, in short, which we associate with living protoplasm.

Because of its dependence upon water, because of the dangers and difficulties which beset land plants, it is easier to imagine the origin of protoplasm in the water; perhaps in the water held in minute channels of the rocks and sands of some early ocean strand, where the temperature and light might be uniform and moderate, where the first faint stirrings of life would not be violently disrupted by the inanimate fury of waves or winds.

It is natural to conceive of the first living things as single units, single cells, which by multiplication might become amorphous masses of cells all alike and all potentially independent; perhaps possessing chlorophyll or some analogous substance which enabled them to absorb energy from the outside world and use it in the construction of organic substances.

Once the mechanism was started, it was autonomous; it

could continue that enlargement and division which was inherent in its nature and become the parent of an ever-widening pool of life. The quickening of the earth was at first only a small element in the processes of nature, little more than a spirit brooding here and there over the waters. But rapidly it increased its range, deposited wastes, died and broke down into soil and so began the formation of that crust of earth on which we live and from which we extract nourishment.

As such masses of green cells gradually extended themselves through the moist places of the earth's surface, carrying on an active life with remarkable powers of indefinite repetition, curious aberrancies must have begun to trouble the succession of generations. New forms appeared, unlike their parents and, when they reproduced, perpetuated themselves. Mutations occur today and we may be sure that they occurred also in those times. New races, new species appeared on earth and multiplied and in turn generated new species.

Through periods of time which it is futile to attempt to measure in years, through long silent epochs compared with which animal life has yet existed but a moment on the earth, the plant world gathered impetus, spread throughout the waters of all the seas, encroached on the shores, experimented with various sorts of form and organization, developed tissues, many-celled bodies which acted as units. Somehow the groups of plants which we know today or of which we have found fossils were roughed out of the primeval vegetation. The first mosses appeared, the first vascular plants and the modern groups of Algae.

Then came the great age sketched above, the age of ferns and fern-like plants and of the great scaly trees; the age also of reptiles. Plants had covered the earth with green; with a bewildering profusion of forms perhaps nearly as numerous as those we know today. Trees like club-mosses or horsetails, tree-ferns, seed-ferns; of these we have palpable evidence. We know much of them and what we know reveals the existence of a large number of very distinct groups, of groups so specialized that they must have been the product of a long prehistory.

Mosses there must have been also and liverworts and Algae in the waters and perhaps colourless Fungi; bacteria must have been present in that mucky soil, or none of the remains of those great trunks and roots would have decayed and we should find the entire debris of the plants and animals of those times. Conifers there certainly were and the fern-like cycads of which we still have living survivors scattered through the tropics. Here and there sprang up some members of a new group, not yet present in great and dominant numbers: the first flowering plants, the first with closed seed-cases.

The flowering plants are the most recent invention of the plant world. Their origin is something of a mystery; their appearance one of the great crises in the history of life. The earliest remains which we can so classify are found in rocks only fifty or sixty million years old. Then, superposed on so many flowerless strata, they appear suddenly in great variety and all over the earth. The oldest fossils of flowering plants are already highly specialized, differentiated and occur in such widely separated places as Texas and New Zealand.

In the rocks of this period we recognize the same families as today clothe the earth; indeed many of the same genera. When we dig into the rocks beneath the living plants we find evidence of a vegetation as varied, complex and highly developed as that which stands there today-sometimes more so. In the part of America, for instance, now called New Jersey, grew three species of tulip-trees, four or five of Sassafras, several of Magnolia, six persimmons, nine bayberries,—even seven species of Eucalyptus. Most of these genera are now represented in eastern North America by fewer species; Eucalyptus by none at all.

Evidently here is a riddle, for the science of the future to solve. Whence came the flowering plants? For it is unthinkable that they appeared so recently as the fossils would indicate and appeared simultaneously all over the earth, already divided into orders, families and genera. There must have been a *first flower* and a long series of precursors of the families of flowering plants found in these recent rocks and still living today. Reasoning backwards from our detailed knowledge of modern plants, we can make a few guesses at the probable nature of these missing links in the chain of life. They were probably woody, perhaps something like the cycads of today. Their flowers were probably large, loosely branching. Fossils are indeed known of cycadlike, fern-like plants which connect ancient ferns with modern conifers and some of these had flower-like reproductive clusters.

The stamens of such early flowers were perhaps branches and themselves branched; the pistils also; the whole structure would suggest a flower-cluster, an inflorescence, rather than a single flower of those we know,-a larger, vaguer, looser arrangement than the highly condensed and arithmetically precise modern version.

Fig. Myosurus

Among living families of plants that which is named for the humble buttercup (Ranunculus) preserves supposedly primitive characters.

Stamens and pistils are indefinitely numerous, spirally disposed on a rather long torus. An extreme example is the little plant called Myosurus, mousetail; the name is for the long slender torus covered with pistils which emerges from the centre of the flower.

Other relatives of the buttercups are larkspurs, columbine, anemone and the gorgeous peony. Rather more distantly connected are the magnolias and tulip-trees. In some of these the pistils are—certainly not primitive—but the simplest known among flowers. Each has the form of a flat and leaf-like branch which has been rolled up lengthwise so that its edges meet and unite. The ovules, attached to these edges, are enclosed in the cavity thus formed, in a single row.

Since those incredibly distant days when the first flowers unfolded to a hazy sky, their progeny have multiplied and diversified, spread far and wide, invaded the waters and climbed towards the eternal snows, settled in the desert and covered the fertile lands with forests.

Some of them have been adopted by man, have varied under his care, have exchanged their old independence for the ability to produce manifold, to transform the salts of the earth and air and water into the fruit and grain and fibre used by him. Others (like the sparrow that hops around the robin and tries to sneak the worm which he cannot dig for himself) have followed man's chosen ones and contend with them in the fields: unlovely weeds, modestly flowered, often ill scented, growing rank and fast.

In all this diversification, which affected every part and which enabled the flowering plants to colonize all parts of the earth; in all this diversity certain features remained relatively stable and enable us to trace probable relationships even in the absence of fossil evidence. The number, the disposition, the fusions of stamens and pistils and petals and sepals give us a clue to those of their ancestors; in the torus we find conductive strands serving these parts in a fairly regular system and sometimes traces of parts that have disappeared—as certain peculiarities of the human circulatory system clearly suggest the ancestral gill-arches. By a glance at some of the commonest living families of flowers we can gain some idea of how the development of the plant world has recently proceeded—and is proceeding.

Whether or not we are right in believing that the first flowers had many and indefinite parts arranged spirally upon an elongated stem-tip, it is certain that most of the flowers of today do not share these characters. If such flowers were indeed the original parents, then their offspring have moved towards flowers with fewer parts, these more definitely arranged in a more elaborate pattern on a cup-like or flattened stem-tip and often joined together in complex fashion. Roses, indeed and the strawberries, blackberries and apples more or less closely related to them, still have the numerous separate stamens and pistils of their hypothetic ancestors.

They have invented a series of modifications of the tip of the flower-stalk, expansions into succulent berries, hips and haws. Individual pistils have in many of them become (in the passage of millions of years) reduced in size, limited to one or two ovules or seeds; the pericarps into which they grow being themselves seed-like rather than comparable with flattened branches or leaves.

Further limitations in the number of parts are seen in that immense brotherhood which may be typified by a sweet pea. The legumes include plants known to Virgil and all farmers since him as valuable for the building of good soil: clover, lucerne, vetches. Also notable foods: peas, beans, lentils. Also great trees, especially in the tropical lands. All these plants are characterized by a striking flower-pattern. Look now at your sweet pea. The petals are not radially symmetric like those of a rose, but are symmetric only about a vertical line.

Above stands a bright "standard"; on either side spread the "wings" and below are two petals joined edge to edge to make the "keel." Force open the keel and you will see a silvery cylinder of ten stamens, all apparently joined together except at their tips. Nine in fact are joined; a little probing with a needle will disclose one, the uppermost, that is not; it springs up with a little pressure, free its whole length.

Press down the sheath formed of the nine united stamens and you see the slender pistil; an almost hair-like green ovary tapering into the upcurved style and stigma. Here is the future pod which will contain the peas—for a sweet pea also forms fruit.

Francis Bacon wrote that "because the breath of flowers is far sweeter in the air (where it comes and goes, like the warbling of music) than in the hand, therefore nothing is more fit for that delight than to know what be the flowers and plants that do best perfume the air."

He recommends further that you set whole alleys of mints that you may enjoy their odor as you tread them. I am reminded of a tent in the Ozark hills which was redolent with the scent of pennyroyal. And was it not written of old: "Why should that man die in whose garden Salvia grows?" Herbs, indeed, were often in demand in former years to mask other odors: the unspeakable effluvia of human living. Now in more hygienic days there may be some question what shall deliver us from the herbs.

These ubiquitous and volatile oils, grateful in the sward if not on thy neighbor's breath, are derived largely from one family; a family which further illustrates the racial tendencies above mentioned and others. The flowers are bilaterally symmetric, all five petals joined to form a two-lipped corolla from which the family takes its name: Labiatae. Stamens, usually four, sometimes only two, frequently in two pairs of unequal size, are mounted on the petals. Two primeval pistils joined to form the compound pistil which occupies the centre of the flower; it is four-lobed in its ovary, two-pronged in its stigma.

Fig. Snapdragon

In the same developmental flow we may find many plants familiar in our gardens, in our medicine, in our commerce. Solanum forms the stem-tuber known to every table as a potato; it also forms small poisonous berries. Eggplants also are Solanum and red peppers nearly related. Lycopersicum, wolf-peach, forms berry of similar structure but edible, called tomatoes. Nicotiana gives its name to the precious drug in its leaves. Strychnine and atropine are produced by its cousins. Snapdragons and phlox are mainstays of our decorative gardens. Not all of these have two lipped corollas and four-lobed ovaries; but all have joined parts, all have a compound pistil; in all the racial movement towards small numbers, condensation, union is apparent.

In certain racial lines another tendency is evident: not only a reduction in number of parts of a flower, but diminution in the size of the entire flower and aggregation of many small flowers into a close cluster. It seems that the development of the original flower from a loose branch-system (if we are right about this) is being repeated by an entire inflorescence. This is well seen in the parsley family, picturesquely called Umbelliferae: their tiny flowers arranged on stalks which radiate from a single point of attachment. In some families a remarkable difference has appeared between flowers of one cluster.

The inner flowers of certain Hydrangeas and Viburnums are minute, not easily found by pollinating insects; the outer ones have conspicuous petals a*nd are sterile*. Certainly here is a sharing of labour which suggests deliberate collaboration or supernatural design, particularly since it is found in several apparently unrelated families. The whole cluster is acting as a single reproductive unit, the outer flowers wholly given over to attraction of pollinators, the inner supplying the pollen and the ovules. Somewhat similar is the "flowering" dogwood (all dogwoods bear flowers), which has a cluster of small yellowish flowers surrounded by large white *bud-scales*.

Pursuing further this line of development among flowers,

we reach perhaps the most successful and ubiquitous of all plant families (other than those of bacteria), called Compositae for reasons that shall be clear. That common weed, beautiful but unloved, the dandelion, is an example. Looking closely at the flower we are at first puzzled; it seems to have no parts like those of other flowers, no central pistil or pistils, no rings of stamens; instead a crowd of petals, becoming smaller as one travels inwards. Each of these petals in fact belongs to a different flower.

The flower of dandelion is not a flower at all, but a group of flowers, an inflorescence. If you bisect it you see a broad disc, on which stand many small buttons; from each of these sprouts one of the yellow seeming-petals; each marks a complete and perfect flower. The basal button may be called torus; the ovary is inside it.

Fig. Helianthus SunFlower

On its summit is a ring of bristles which answers for a

calyx. Within this is the corolla, of five united petals, tubular at the base, but in its upper part slit down one side and flattened into the strap-shaped yellow "petal" first noticed. Through the tube of the corolla projects a ring of five stamens, all joined, forming a slender cylinder or sleeve. Inside this is the style, two-pronged at the tip.

All the Compositae have this general plan. It is easy to realise how common and successful they have become when one recites a few of their names. Daisies, asters, goldenrods, zinnias, chrysanthemums, sunflowers, thistles, lettuce, ragweed... we might continue to enumerate garden beauties and common vegetables and calamitous weeds which belong to this family.

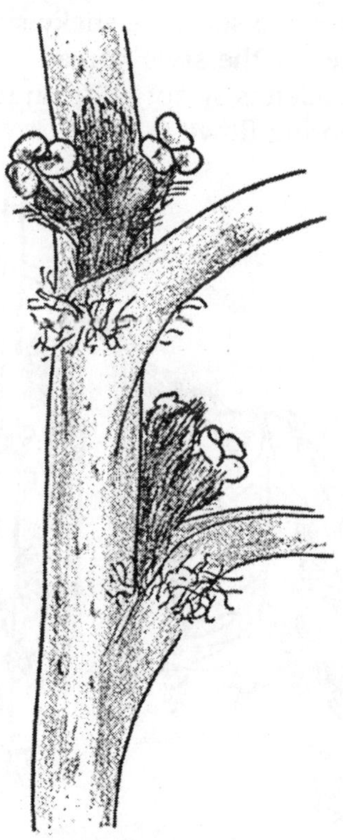

Fig. White Oak

Many are more elaborate than dandelion; they display two

sorts of flowers in one "head." Around the edge are flowers like those described above, with strap-shaped corolla. Within this circle are many smaller flowers, with tubular symmetric corolla. In Helianthus—sunflower—and other Compositae, only these disc flowers are perfect, possess stamens and pistil. The rays are sterile, serving only for display; another version of the arrangement noted in Viburnum and Cornus.

The stamens open to the inside, so that the pollen is enclosed in that small sleeve formed of the united anthers. At this time the style has not yet matured. Now it grows longer, pushes up through the anther-sleeve, carrying the mass of pollen on its tip out of the corolla.

It is not itself pollinated thereby, for its two prongs are still folded together, the sensitive sticky surfaces unexposed. When the branches of the style spread apart, the stigma is exposed and the pollen is swept aside on to the already open stigmas of neighboring flowers.

Fig. Quercus

Lest the reader suppose that we have reached at once the

end of this line and of variation in plants;—let him glance at a few flowering trees. By this is meant oaks, hickories, willows, elms. These have flowers indeed, descended from the same ancestors that fathered roses and snapdragons and dahlias; but descended by a process of reduction to essentials, a simplification. Insect pollination is abandoned, the pollen takes to the air.

The catkins of an oak, which hang from the twigs in spring and later litter the walks, are long lax clusters of small flowers. Each flower contains, within a green perianth, stamens but no pistil. In the axils of the leaves, somewhat higher on the same twig, are inconspicuous flowers which consist only of pistils surrounded by a ring of scales. The pistil becomes the acorn—a pericarp; a structure of apparent simplicity which contains the evidence of long history, of a larger and more complex flower in its ancestry.

Maples have flowers which often have both a ring of stamens and a pistil in the middle; or, in some species, pistillate flowers on some trees, staminate on others; or various combinations of these conditions. The pistil is a beautiful two-lobed, two-branched object, which later becomes the double "key" so easily dispersed by the winds. Willows and poplars have all their flowers in catkins, each flower a pistil or a cluster of stamens, each partly shielded by a small scale which bears a nectary. These scales, in some kinds of willows, bear long silky hairs and before they are fully expanded form the "pussy willows" of early spring.

One great tribe of trees is set off from all these; it has no proper flowers. Here are the evergreens, the cone-bearers, the pines and cedars and spruces and firs and all their kin. A great and widespread group of plants, crowning mountains all over the earth, ranging far to the north in the trackless wilderness, covering the hot moist lowlands of the south. Their leaves are needles or scales. The whole being of some is permeated by tubes which contain resin, that sweet-smelling sticky exudation which man has put to many uses without discovering how it serves the plant which produces it.

Instead of flowers these trees form cones. Early in the

spring appear the delicate soft clusters of overlapping scales, from which pollen sifts out; yellow pollen in visible clouds. These cones, having shed their dust, wither and fall. Others, still smaller, receive some of the pollen. On their scales are ovules, which are pollinated and become seeds; all going much as it goes in flowers.

The cones become larger and harder, the familiar cones which everyone has seen. These are the equivalents of the pericarps of flowers; but each scale bears the seed on its surface, not in an interior cavity as a pistil, a proper pericarp, does. The cones are undoubtedly the condensed product of simplification of an older and larger branch-system, perhaps like that which was the parent also of flowers.

Other great families, in no way related to the trees above sketched, yet exhibit the same tendencies towards simplification and reduction of the flowers. One of these is a group which has undoubtedly had more influence on the lives and fortunes of men than any other one family of plants, perhaps more than all others together. The history of man has been a process of settling down; a change from a nomadic way of life to a static; from hunter to shepherd, then to tiller of soil.

The rise of agriculture meant the growth of the clan, the development of communities, of society, of government. Of all the plants that have participated in this evolution, of all the plants sown and reaped by man, the cereal grains-wheat, barley, oats, rice—are the most ancient and the most important. So widespread has their use been that we learn something of human history and wanderings from the grains which they have used.

The history of cultivated wheat goes back so far that we have scarcely a trace of its wild ancestors and cannot certainly name the country of its origin. A Russian botanist has given much time to charting the variability of the races of wheat in relation to geography, on the supposition that in the centre of greatest variation lies its origin and perhaps also the origin of human civilization.

Another botanist has pointed out the fallacy of the old myth of Atlantis, the lost continent, or other fabled means of communication between the old and new worlds accessible to civilized man. For the inhabitants of the Americas had not a single grain in common with those of the rest of the world; they were innocent of wheat, as the others were of maize. They must have come as nomads, hunters and developed their own culture independently. Their only domestic animal was the dog; the companion of wandering hunters to this day.

Another section of this great family has played an almost equal part: the bamboos. We of the western culture know them chiefly as ornaments in botanic gardens and as materials for fishing poles. But there are great expanses of land thickly populated by men to whom the bamboos are the universal and all-important timber.

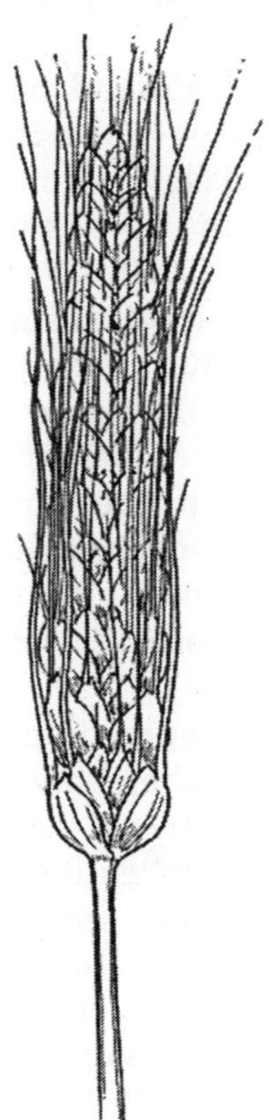

Not only are the young sprouts eaten in "chow mein."

The hollow jointed stems are cut into sections, made into pipes and buckets and all sorts of vessels. Fibers become mats and cloth and rope. The light firm uniform sticks determine the style of the architecture (which always precedes architects); light frameworks of slender poles supporting a thatch of bamboo straw. If our own civilization is based on wheat and its relatives, so that of southeastern Asia might be spoken of as a bamboo culture.

Grasses, like oaks and poplars, are not classed among the flowering plants by the uninformed. It is true they form seed; but the idea of flowers on a grass seems a contradiction. In fact the flowers are small, greenish; none the less complicated and effective. A microscopic view opens a new world of beauty.

These flowers are arranged in spikelets, themselves grouped in the dense spikes or the graceful feathery panicles which tip the stalks. A spikelet is formed of some exterior green scales, within which are the flowers (sometimes only one). A single flower is enclosed first by a scale named the lemma; this embraces a second scale, often smaller and more delicate, named the palea. This in turn is more or less folded around the stamens and the pistil. The stamens are pushed out, when the flowers open, by the lengthening of their slender stalks; the anthers hang loosely in the wind, thousands of them, shedding their pollen; these delicate things impart the feathery appearance to a spike of orchard grass or a panicle of bluegrass; those of timothy can transform the entire field, cover it with a tenuous purplish bloom.

The pistil boasts two spreading styles, covered with a profusion of minute branches, any of which can serve as stigma. These also project between the enveloping scales, easily catch some of the drifting pollen from the air. In the pistil is one ovule. The ovule becomes seed, the ovary matures into pericarp. The peculiarity of this fruit is that the mature seed is inseparable from the pericarp around it; the two have grown together. This is the grain, the staff of life to many millions of men; neither a seed nor a fruit in the usual senses of those words, but both, a one-seeded fruit.

Here you have had a few glances at the many-coloured, many-sided, many-faced world of the flowering plants. To provide an adequate description of the group would necessitate many more pages, a book, several volumes. We should speak of the heaths, of the palms, of the sedges, of the gorgeous orchids and many others. But the scant sample provided has been selected so as to illustrate certain tendencies which have appeared during the existence of the group, certain changes which have manifested themselves.

Fig. Flowers of Wheat

As the flowering plants conquered the earth and diversified themselves into the most varied of all groups of plants, their flowers became smaller, having parts limited to exact numbers and often united with each other. As flowers lost stature they become increasingly social, gathered into inflorescences which themselves seem like single flowers. While in some races corollas were formed of increasing complexity and differentiation, in others the corolla disappeared altogether, the flowers abandoned all efforts at display and embarked on wind pollination instead of begging help from the insect world.

It is well to note also that these changes may correctly be spoken of as tendencies; there is indubitable evidence that some of them have been manifest in quite independently changing lines of descent. The similarity in many respects of a single flower of grass or sedge with a single flower of dandelion or sunflower is a coincidence, a product of unrelated races; so is the inflorescence of a Viburnum with that of a Hydrangea. In the shifting scintillating kaleidoscope of the flowering world there seem to emerge a few definite patterns of variation.

Fig. Orchid

So much is history. Parts of the chronicle are fully documented; we may be as certain of them as we are of the campaigns fought in Gaul under the leadership of Julius Caesar, of the development of parliamentary government in England. Parts are dim or entirely obscure; we supply these by inference and analogy as we fill in the details of the earliest periods of the Egyptian kingship or of Europe under the Germanic hordes which destroyed the Roman Empire.

Even when good evidence is lacking, we may be fairly confident that our carefully considered opinions contain some glimmering truth. The plant world has a history, infinitely longer and infinitely vaster in its scope than the petty chronicles of man. We know some of this history in considerable detail and much more of it in general outline. Our knowledge is recent, is constantly developing and will without doubt continue its rapid expansion into the unknown.

Animals also are involved in the long story of the world. They seem to have arrived on the scene later; they were almost from the beginning dependent upon plants; they varied, as plants varied. They have a special interest for us because of our own animality. Biologically we count ourselves the latest products—some say the "highest"—of this vast industry of animal production. For this reason, as we contemplate the history of the earth we are not content merely to pass the facts in dispassionate review. We create a philosophy wherewith to describe it, to "explain" it; if history has resulted in humanity, it must (we think) have some significance for humans.

"Let sleeping dogs lie," wrote Jefferies, "and evolution is a very weary dog." In fact it has been kicked about. The speculative biologist (and in former times he was in good standing among his fellows) early seized upon it and built it into his conception of the universe. In his hands it became embellished with authority, dignified with scientific prestige.

The Chevalier de Lamarck, brilliant French naturalist, developed a "zoological philosophy," in which were expounded the laws of variation. His ideas were not mechanical, in the modern sense, for they invoked as causes the "needs" of organisms, which progressed in their organization because of the "forces" within themselves; progressed from vague and generalized beginnings to the marvelous adaptions of the bee to the flower and the bird to the buoyant air. However, though conceived in terms of purposes, it was a naturalistic philosophy.

Others identified evolution with various supernaturalistic schemes which had come down from ancient days. To Aristotle, the father of science, the whole of animal creation demonstrated an ideal series, a progression towards perfection, an abstract order of nature. To medieval churchmen this became a plan in the Divine Mind. Saint Augustine has been accused of a belief in evolution, in some modern sense or other;

it would have shocked that blessed old ascetic to hear his vision identified with a purely mechanical sequence of cause and effect. There is little evidence that any thinker in medieval Europe thought of the forms of life as *successive*, as appearing by natural causes through a sequence of time.

Modern evolution theory is naturalistic in describing the development of life through the ages as due to natural agencies. It envisions a primeval world in which life was not; then the beginning of life and its slow transformation through the unending series of generations, *one form giving rise to another* in accord with definite laws. Life unrolls, evolves, as the earth spins. Evolution really contravenes heredity. Living organisms seem to inherit from their parents that which makes them hold fast to their type; but ever within them there is that which constantly makes for unlikeness, for the development of the new, for progression towards the ideal or towards an appointed end.

These comprehensive and sometimes grandiose concepts which I have vaguely adumbrated are not, strictly speaking, scientific; they have a doubtful place in the science of biology. It is true that biologists have played with them, adopted them, written about them and unendingly confused them with their more scientific statements; but biologists are human and must be allowed their diversions. Almost in the same breath they tell us that new species have "evolved" through a hit-and-miss random assortment of mostly valueless characters and that evolution has steadily progressed towards the "higher" forms, culminating in man.

In another scientific pasture we see eminent physicists gravely but somewhat incompetently venturing down seductive metaphysical lanes into discussions of the nature of God. Which is well enough, providing we are not led to attach to their conclusions about the Almighty the same credence that is owing to their strictly scientific generalizations. Let us always be careful to separate from the physics of a physicist his religion, his golf and other facets of his human nature. So also let us distinguish from evolution theory in the classical sense—or senses—the scientific study of the origin of species.

The older naturalists were little concerned with this question. "So many species are there," said Linnaeus, "as the Infinite Being in the beginning created diverse forms." Variation was considered to be due only to the environment. Occasional hybridization might produce new forms, but these were sterile, left no progeny. Species were immutable, perpetual. It did not occur to his immediate followers to question this attitude, though it entailed some vigorous disputes on the "reality" of species in the more complex (and, as we now know, more rapidly changing) groups.

If they thought of evolution at all, it remained a purely philosophic speculation, having to do with the plan of things, with that natural order which they considered it their duty to reveal, but not with the actual production of changes in living form. Whatever their divinely ordained relations with each other, whaever their significance in the order of creation, the species spawned their way on through the ages as they had done from the moment of their origin.

As every schoolboy knows, the end to this peacefully static world of biology came in the middle of the nineteenth century through the work of Charles Darwin. In 1859 he published his remarkable book, entitled *On the Origin of Species by Means of Natural Selection or the Preservation of Favoured Races in the Struggle for Life*. This was the first scientific attempt to survey in comprehensive fashion the facts of paleontology, of heredity, of variation, of relationship and to draw therefrom the inevitable conclusion; moreover, it contained the first considerable attempt to provide a purely scientific, mechanistic hypothesis to account for the origin of new species.

Darwin was a keen and persistent observer, a careful and thorough thinker, a man with a vast store of information and a noble power of synthesis. He knew his statement would shock biologists and others, all unprepared for a concept of transmutation of species, of "modification by descent" (to use his own oft-repeated phrase). It is significant that he did not refer to his theory as evolution—a speculation with which he was not concerned. Prophetically enough, the last word in his book is "evolved."

Prophetically enough, because in the ensuing storm Darwin's theory was taken up by the evolutionists and identified with their own. Conversely, evolution was taken up by the scientists; teachers, mostly uncritical as they are, took the doctrine to their hearts, clothed it in the garments of natural selection and through the outcry of outraged religious and romantic sentiment championed it as a scientific doctrine.

Thomas Huxley, the great teacher, expounded good Darwinian science under the name evolution. From his vigorous and lucid exposition it fell into the hands of Herbert Spencer, romantic and slipshod natural philosopher. By him Darwinism was used as the keystone of a philosophic arch through which the teachers of biology, mistaking it for science, felt compelled by sheer weight of eloquence to enter.

They came to regard evolution as the focus of their subject, without which it had no meaning and indeed could not be taught. A student of living organisms might become not only a professor and the chairman of a department but a defender of the faith. The seeker after nature must now venture warily into a horrid tangle of beliefs, prejudices, theories, facts, names and vested interests. He must expect to find many different things called evolution according as he deals with botanists, zoologists, sociologists, psychologists, theologists and plain parsons.

So much of a digression has been thought necessary to put the reader on his guard. Let us here have done with evolution; with progression from "lower" to "higher" (whatever that may mean); with the appointed aims and ends and ideals of life and return to a purely scientific consideration of the origin of species, of the causes of the diversification of the living world briefly sketched in the preceding pages. Let us admit before we begin that science has not yet found any very satisfying answers to the problems posed by such a study. Answers have been proposed; some have been proven false; considerable knowledge has been acquired; research holds the promise of increase of knowledge.

Perhaps it is worth repeating that there can be no doubt that new species do actually arise. Darwin advanced a sufficiency of evidence, mostly indirect; since his time the conclusion has been reinforced by an abundance of direct evidence which leaves no hesitation in any reasonable mind.

One of the most evident facts of nature is the instability of heredity, already encountered in a previous chapter. The chips from the old block are not all alike, even in the same environment. Plants and animals do not always breed true.

The often delayed results of cross-breeding cause racial divergences. Actual changes seem to occur in the inherited protoplasm itself; new genes replace old ones or are added to them; chromosomes are taken apart and joined in new patterns. As surely as individuals reproduce new individuals, so surely do races give rise to new races. Our agriculture has always recognized this; every farmer who can accomplish more than bare subsistence has striven to "improve" his plants and animals, to change the varieties which he has into new ones. In fact, all our domestic creatures have originated in this way from wild plants and animals, some of them within the small space chronicled in written history.

The tardy recognition of the real significance of variability, which has been known as long as crops have been cultivated and cattle reared, was partly due to the influence of Linnaeus, whose system of classification shaped the course of biological history for a century and whose opinions were received with the reverence due to a master. Modern research, however, has shown that many of the slight differences to be observed on every side in nature and under cultivation are really hereditary; that what Linnaeus thought of as a fixed species somewhat diversified by the environment is more often a collection of many similar and often interbreeding but none the less genetically different races.

The origin of races is an observed fact, not a debatable theory. The origin of particular varieties of sweet peas, potatoes, raspberries, roses, sheep, cattle has been observed and recorded. There is no reason to impugn the veracity of those who have reported such changes. Furthermore, hereditary variation may be accelerated by experimental means; scientists of today can bring new races into being at will by using such tools as X-rays (they cannot always predict what their new characters will be).

There are those who would save what they deem to be some necessary fortress of religion or skepticism by admitting these facts but denying that the new things which appear are ever species or can ever lead to the appearance of species. They distinguish between varieties, which may originate in the manner just mentioned and the divinely instituted species which include such varieties and from which no variety can diverge enough to be excluded.

They adhere to the distinction made by Linnaeus, that varieties within one species will interbreed but species will not (to give fertile offspring). These persons, whose arguments may seem at first sight plausible, simply do not know plants and animals. Everyone who has worked long and thoughtfully with the classification of living things in recent years knows that the lines between many of our species are purely arbitrary and do not represent a real cleavage in nature.

They know that species which occupy distinct geographical ranges lose their distinctness when, through artifice or cataclysm, they are mingled. They know that hundreds of species, recognizable in the field by any qualified person, pass through series of hybrids, at the fringes of their ranges, into neighboring species. They know that if the Linnaean criterion were applied, all the common blue stemless violets of the United States would have to go in one vast species—for they all interbreed.

Finally it has been possible through artificial hybridization and other laboratory techniques to create new species which cannot breed with their parents but which will reproduce themselves. These synthetic species have actually duplicated existing wild species, leaving little doubt that the latter had originated from the same parents in the same way.

What one biologist calls a species another may call a variety and vice versa. In describing a new form the scientist frequently presents his reasons for considering it a variety rather than a species; reasons sometimes purely of convenience. The races of which species are composed being variable and often isolated either by sterility or by outer circumstances from each other, the species in which they are included are obviously but temporary and unstable groups of races, which must ultimately split into smaller species.

It is precisely for this reason that there are genera in nature and that the genera are arrangeable in families, the families in orders and the orders in classes. The similarity of the species of a genus is due to their relationship; they were once varieties in a species. The admission of the fundamental mechanism of variation in plants and animals concedes the origin of new species. Darwin perceived this very clearly and devoted pages to the demonstration that varieties are incipient species.

Some day, perhaps, some one will write a history of the theories of origin by descent. There were some attempts at explanation even before Darwin (always aside from evolution theory). But Darwin's was the first entitled to serious scientific consideration; it towers above all others.

It must be borne in mind that a theory of descent—a theory which explains the *causes* of the appearance of new species— must explain other things besides mere novelty; it must explain why we have *these* races rather than others. All kinds of living things are, in various degrees, adapted to the conditions under which they live and reproduce. The means of adaptation are often, to our naive intellects, astonishing.

We want to know how the leopard got its spots and the gazelle its speed; what gave the elephant its trunk and the kangaroo its pouch; how a humming bird acquired the long bill which sips the nectar from flowers and how flowers came by nectar and the long corolla-tubes in which it collects; how the first leaf was made and spread a flat surface to the sun and how its minute air-holes and air-chambers, its watercarrying and food-carrying tubes were first formed; how that aeronautic miracle, a feather, was invented; how toebones got changed into suitable frameworks for feathered wings, or into leaping or tearing or climbing or violin-playing instruments; how the supposed original naked jelly-like living matter grew hair and hide, eyes, nose, ears, hands and feet and wings, leaves, roots, pollen and pistil and seized as its covenanted land the whole expanse of the earth and the waters thereof.

The concern with adaptation led Lamarck into purely teleological statements and an invalid postulate of heredity. The failure of modern science to demonstrate any effect upon the next generation of the experience of this has already been mentioned. Though Darwin avoided such pitfalls, the concern with adaptation was ever with him and shaped his theory as a stencil shapes the letters on a signboard. The outlines of the Darwinian theory, the theory by which he sought to account for the origin of species, have been more or less competently expounded in scores of textbooks.

Strangely enough the theory itself is not really well known. Few students who have "had" biology or botany or zoology can explain what Darwin proposed and I have known Doctors of Philosophy in these sciences who had never read a word written by the great biologist. We shall not do amiss to review the theory here—if only to discard it later.

Darwin's theory begins by pointing out an obvious fact-a good way for any theory to begin. This fact is that the world is continually being furnished with a greater number of new individuals than can possibly survive. The prodigality of nature in the matter of offspring is always a shock to those whose knowledge is limited to the more superficial aspects of human procreation.

An ordinary mullein, a large woolly-leaved plant which grows on rocky hillsides and dry pastures, may form a quarter of a million seeds. If we allow a square foot for each plant, one mullein could cover nearly six acres with the close ranks of its progeny and if this were repeated the following season, the descendants of the original plant would occupy more than 2000 square miles. But this is moderation.

The possibilities of a fern have already been described in these pages. If we turn to the so-called "lower" plants, the figures become fantastic. Though a bacterium is so tiny that it is difficult to express its size in cubic inches, yet the progeny of one individual after 24 hours (supposing that reproduction occurs, as it frequently does, every 30 minutes) would make a mass of nearly 300 cubic inches; about a gallon of bacteria. What the numbers would be after another 24 hours staggers the imagination. The progeny of the single original microbe could then cover all England to a depth of about a foot and one-half; or Pennsylvania to a slightly greater depth.

There are many mulleins on a hillside that I know and I expect that there will be no more of them many years hence. There may be three hundred million bacteria in a quart of good milk; but I do not anticipate being buried alive by bacteria within a few weeks. The great majority of all the living things engendered never reach maturity, never even begin growth. They may lie dormant for a while, or they may die at once. Only a minute fraction of all the millions mature and take the place of those that pass away.

A newly deposited mud-flat in a big river may become quickly coveted with seedling willows, as many as 250 to a square meter. After a year or two you will find only about 50 survivors in each square meter and, if the whole colony escapes being swept away by the spring floods, two or three individuals of every original group of 250 may form the willow thicket which eventually stands there.

The survivors must often owe their survival to small advantages which they possess; it cannot be always entirely a matter of chance which individuals die, which live. If some young willow shoots are faster-growing than their neighbors they are more likely to flourish, to overtop their weaker sisters, to receive more than their share of light and air. If they have longer roots, they may steal some of the water from the plants around them.

Likewise if any of a litter of fox-pups survives it is likely to be the fleetest and most resourceful. In short, if there are any differences among seedlings, sporelings, fledglings, cubs, larvae (and there always are) and if these differences are of any importance in securing food or eluding enemies; then it is obvious that the few survivors are likely to differ from the many that perish by just these differences. In the modern phrase, their individual traits have *survival value*. This contention occupies a good part of the *Origin of Species* and is there supported by an abundance of evidence. Notice the clear differentiation between such a theory and Lamarckian theory: Life is a result of adaptation, rather than something which makes adaptation desirable. Individuals survive because they happen to be adapted and most die because they are not; plants and animals do not (on this theory) adapt themselves *in order* to survive.

The final link—and the weakest—in the Darwinian theory is the assumption that the characters, the peculiarities responsible for the survival of certain individuals, will reappear in the progeny of these survivors, at least to an increased extent. Because of the premium put upon swiftness in deer, an entire generation may be thought of as the offspring of the swiftest individuals of the preceding generation, who alone matured and reproduced; *hence the swiftness of the race.* Because only the faster-growing plants survive and form seeds, their offspring represent a higher level of rate of growth than did the previous generation; they inherited the ability to grow fast.

If of all the seeds that fall upon barren ground, upon exposed rocks, a few, through unusual thickness of cutin or other protective features, are able to germinate, grow and establish themselves as rock-creepers; then a new race of lithophytes, perhaps even in time a new species, may arise, marked by those traits (intensified by the additions of later generations) which enabled the first pioneers to grow.

Such is the origin of species by natural selection. It is to a large extent an attempt to explain adaptation. The plants and animals which win in the struggle for life are those which are best adapted to something or other. It is worth remarking at this point that adaptation has been somewhat overworked. The presence of the bird in the bush is not an example of a wonderful sagacity which is the climax of long generations of selection; it is the inevitable result of the size and habits of the bird. It is clever of birds to build nests; but it would be greatly to their advantage to devise something better, from which the fledglings would not fall so easily a prey to four-legged hunters.

Of the thousands of seeds carried on a flood, the one that happens to swirl on a particular square centimeter of water may happen to touch land; its neighbour, with just as good qualities or better, is less favored by fortune. In short, chance is a factor in survival. But, even after we have discounted romantic exaggeration, we recognize that life is, after all, an interaction between protoplasm and environment. It continues by virtue of the circumstances with which it is correlated, which favour some individuals, eliminate others.

Modern exponents of evolution have often overlooked something which was plainly stated by Darwin. Is not an ameba perfectly adapted to its environment? Is a man perfectly adapted to his? There is no need to fill more books with descriptions of the intricacy and perfection of the human eye, the nervous system, the digestive reactions and so forth; when you are next tempted to brag of man's physical perfection—think of your dentist. Is it not a sign of superiority in an ameba that it can get along without such imperfect and troublesome gadgets as teeth? It is difficult to understand how any new species could get started merely through an increasing degree of adaptation to the environment; its parents already being well adapted.

The virtue of variation is, as Darwin saw so clearly, that one may thus obtain adaptation to a *new* environment; it provides a chance of escape from an already crowded microcosm. If the offspring of an alga, which makes its own food by photosynthesis, can through a heritable change begin to use as food the organic remnants which may fall into the pool, both the new type and the old can exist side by side without mutual hindrance; otherwise parent and offspring must compete for the light and carbon dioxide in the water. Diversification means economy, greater use of natural resources.

A limited body of water or patch of soil or rock can support more life if this embraces several kinds of plants and animals, differently adapted to the various conditions there. All the niches of all the environments must be filled before we need fear a blocking of new species through a lack of opportunity for better adaptation. Through individual adjustments and the deliberate modification of his surroundings, man has been particularly successful in using various environments, without becoming several new species in the process.

He is indeed not very well adapted to any natural environment, except as he has been able to change things, to fell and build and plough and tunnel and bridge and sail and fly; by means of his inventive genius he has managed to fill a greater variety of environments than any other living species.

But in expounding adaptation we are apt to lose sight of the main problem of the origin of species; Darwin himself did so. The fascination of showing how *these species* have been favored in the race for survival has led us away from the need of showing how new species of *any kind* can have arisen. The origin of something by selection is perilously close to a contradiction of terms. Natural selection is really natural rejection; the survival of the fittest is the elimination of the unfit.

When we recall the poverty of Darwin's knowledge of heredity we can understand why it did not seem to him highly important to discriminate between the origin of variations and the survival of new species; or even to enquire into the causes of the variations themselves.

Variations are self-evident and it was taken for granted that they would reappear in the next generation. As through most of recorded history the "transmission" of a mutilation to the next generation was accepted without skepticism, so it was generally believed, without criticism, that the tallest trees have the tallest seedlings and the most powerful panthers the most powerful offspring.

We know better today; we know that some differences are due indeed to differences in the inherited germplasm, but most are to be accounted for by small and often imperceptible differences in environment. In spite of this knowledge, the old manner of speaking lingers and inspires many a careless and misleading statement about the origin of species.

Few of those who expound Darwinism mention the classic

experiment which terminated Darwin's principal assumption. It was an experiment with beans; a simple experiment, as farreaching in its implications as Mendel's with peas. It was performed by Wilhelm Ludwig Johannsen in Denmark and published only forty years ago. Starting with the observation that beans even in one variety, even in one pod, differ in weight, he asked: Do the plants grown from the heavier beans form beans heavier than the rest?

Beans are self-pollinated; it was possible to begin the experiment with a single bean plant and the beans which it formed; to grow each of these seeds and harvest the seeds from the resulting plant and so on generation after generation. The answer was clear: the average weight of the beans produced by any plant had no relation to the weight of the bean from which that plant grew, but was always close to the racial average.

By selection of heavy beans from each successive crop absolutely no progress was made in obtaining a heavier variety of beans. The differences in weight were doubtless the results of minute differences in the position or nutrition of the beans, differences which affected them while they grew and which can be classed as environmental. The heredity of all was the same and did not change during the experiment.

Similar results have been obtained with many other kinds of plants and by many breeders of plants. Some, it is true, have apparently been able to "create" improved varieties of plants and animals merely by selecting the best every generation for breeding and discarding the rest. *But only when the race had a mixed heredity*. Most domestic crops (and animals) are vast and intricate mixtures of heredities, containing in every cell a jumble of chromosomes and genes of different ancestries.

When you select the best individuals from such a population, you are selecting those which contain the best genes. When you mate them together to obtain the next generation, you are assisting in the formation of a race in which the desirable genes are relatively pure.

Finally you will have a group with a more or less pure heredity, a true-breeding variety which contains the desired improvements. Further than this you cannot go by selection alone. Neither human nor natural selection *creates* anything new in the heredity. Although they have officially discarded Lamarckian evolutionism, those who teach evolution often slip back into a Lamarckian point of view.

A change occurs because it is useful, then is preserved by natural selection. A flower develops a nectary in order to attract insects; nectary-possessing flowers are favored in the struggle for life and so perpetuate their peculiarity. Darwin evaded the problem because he did not realise its difficulty. The honest scientist must now say that the flower succeeds because it has a nectary, but *we do not know* the causes which induced the formation of the first nectary.

The causes must precede the effects; we must look in the past history of the nectarless flower for some cause which changed its development, caused certain cells to fashion themselves into a small succulent mass and to exude a sweet secretion.

Whatever the cause was, it must have been some change in the reproductive cells which formed the entire flower, pollen and ovules as well as nectaries; otherwise it would not reappear in the next generation. The origin of species does not occur in the mature parts of individuals but in the meristic cells that lead to the succeeding generations.

The nectaries and petals and anthers, the claws and hooves and wings and fins, the leaves and feathers and eyes and hands and tendrils when they first appear are the visible signs of changes that have already occurred, occurred in the quiet and secret cells of embryonic tissue. We have a name for these changes; of their causes we are as yet ignorant.

The name for hereditary changes in the germplasm is mutations. The word was popularized by the botanist Hugo de Vries, who used it to describe what happened in his garden in Amsterdam. In the progeny of certain evening primroses (Oenothera) appeared new types which at once behaved like distinct species, breeding true and themselves often "throwing" new types.

Here were species produced all at a bound, by "saltation,"

not by the gradual accumulation of minute differences described by Darwin. De Vries supposed that he had stumbled upon the origin of species in the process.

Later study has shown that these mutations in Oenothera, though undoubtedly hereditary, are really new combinations of germinal materials rather than the production of any new matter. The parents were really hybrids of an extraordinarily complex kind; because of unusual factors they might seem to be a true-breeding race most of the time, but now and then the restraining influences were evaded and the hidden elements in their composition found expression.

Since it appears that de Vries' mutations are not really anything racially new, his name has been transferred to what he thought he was witnessing: changes in germplasm. If a gene changes or if a chromosome changes so as to influence development in some new way and so produce a different race, the change is called a mutation. So far from being the "saltations" described by the Dutch botanist, the mutations discovered in our laboratories have turned out to be usually very minute changes indeed, often barely perceptible even to the expert.

The search for the causes of these changes is one of the most active fields in modern research; a field which is advancing so rapidly that any general statements here made would be almost certainly out of date within a year or two. We know that new races originate through the rearrangement within cells of chromosomes or fragments of chromosomes and through changes in the number of chromosomes.

It has been possible to produce some such changes experimentally, to bombard reproductive cells with X-rays and other types of rays and so to cause pieces of chromosomes to detach themselves and to fasten themselves on again in new places or with their ends reversed. It is becoming increasingly evident that many species differ in having multiple sets of chromosomes; the parent species having two sets in the normal way, some of its derivatives may have four or six sets or even more.

Since some of these may breed together (though others

will not), the resulting picture rapidly becomes extremely complicated, with races differing in all sorts of numerical relations.

There is no doubt that such changes (of whose causes in nature we are still largely ignorant) have played a large part in the speciation of populations of plants and animals. The multiplication of sets of chromosomes is particularly prevalent among the flowering plants and has played a large part in the development of their innumerable species, their versatility, their rapid conquest of the earth.

Such changes having been duly considered, there remain apparently others which at present are simply labeled "gene changes." Genes seem to change without appreciable change in the chromosomes which bear them. Wherever a gene changes there is some slight or great change in the development of the organism, ending either in failure to live (many mutations have this sad result) or in the appearance of a new "character."

The modern geneticist has come to regard the origin of species as brought about by a series of minute steps, changes of all kinds, often of no particular value, which occur comparatively rarely in our short perspective of time, but which none the less occur.

As the germinal materials of related plants and animals become diversified by such changes, through hybridity and multiplication of chromosomes the number of combinations of differing heredities is greatly increased and many viable and successful varieties and species result.

Natural selection here exerts the influence expounded by Darwin and the modern view is in fact a sort of rejuvenated Darwinism, "neo-Darwinism." It differs from classic Darwinism in that it envisions variation as something sporadic, affecting only scattered individuals which must compete with the much more numerous parental population.

Darwin thought of variation as a constant process affecting large segments of a population in the same way at the same time, so that an entire species could change into several new species and leave no trace of the old.

It remains for future science to discover whether genic inconstancy and chromosomal acrobatics are indeed sufficient to account for the history and the future of the plant and animal worlds; or whether they are simply frills on the skirts of dame Nature who has so far remained decently veiled. It remains for future science to discover whether the origin of species is by a sort of flea-jumping hither and thither or by a steady march in a few well established directions.

The fortuity of the variations puts a severe strain upon the faculties of the most credulous.

We have all geologic time at our disposal; even that may seem insufficient. It has been said that a battalion of monkeys with typewriters could merely by their aimless pounding on the keys duplicate the entire literature of the world—*given enough time.*

With a similar qualification we can perhaps envision any such nonsensical process; is the origin of species to be so characterized?

Fig. Orchid Family

It is a pity to leave this subject in such an unsettled state. The origin of species is being attacked by more ability and more kinds of technique than have ever before been mustered; but the answer is not yet.

Let us not lose hope and gently sink back into the sheltering arms of evolution theory.

The origin of species is not intrinsically insoluble. Some day—if our civilization does not destroy itself—the curious mind of man, working in the beautiful disciplines which it has created, will discover the moving forces which cause variation in germplasm, which shape the species of the future and will chart their nature and their mode of action.

Perhaps this will reconcile scientific generalization with some evolution theory or other; which, we must not now dare to predict.

Chapter 10

The end of Things

The leaves fall gently to earth; their bright colours fade, their substance withers, they cease from compounding water and air into food: they are dead. Then the final act of the comedy begins. Life is a prelude to death; death is the threshold of dissolution. The hand of nature is laid softly on her fallen children and they are received back again into the earth from which they sprang. Even the semblance of organic form gradually vanishes, as to dust they return.

The processes of decay, different as they seem from those of life, usually offensive to our educated senses, are generally regarded as an unpleasant and regrettable epilogue to life, to be avoided in pleasant discourse. There may even be sometimes a suspicion that the corruption of dead things is a flaw in the perfection of our world, which in the best of all possible worlds would not occur. Certainly the majority of mankind have been at considerable pains to postpone or even to wholly avoid the natural dissolution of their own bodies.

To a well balanced philosophy such a prejudice seems regrettable. The decay of bodies is a link in the chain of life. "How true," wrote Thomas Carlyle, "that there is nothing dead in this Universe; that what we call dead is only changed, its forces working in inverse order! 'The leaf that lies rotting in moist winds,' says one, 'has still force; else how could it—*rot*?'"

There is sound biology in this thought; there is also an assumption that is open to question. To suppose that decay is a sort of inversion of life is a view so old that age alone entitles it to veneration, even though it turns out to be false. So universal, so familiar is decay that the first scientists took it for a natural and inevitable characteristic of organic matter. As they came to understand something of the forces and processes that operate in living matter, they held that decay was the result of the cessation of these activities and the supervention of new activities equally natural to protoplasm. There were, in fact, abundant signs that decay is closely connected with life.

Do not mushrooms spring from decaying piles of manure or from rotting leaves? Is not their life generated by the death and dissolution of the substance from which they spring? So bread becomes covered with mold, so the juice of grapes ferments and so, it was held, maggots originate in spoiling meat. On the grave of the murdered maiden a violet grew and miraculously the human soul was transfused into the flower.

Somehow it occurred to an inquisitive lawyer, named *Francesco Redi,* who lived at Padua in the seventeenth century, to doubt the common belief about the origin of maggots. It occurred to him also to put the matter to the test of a simple experiment. In doing this he was manifesting what we should call a modern scientific attitude; most of his contemporaries, being still medieval in intellectual outlook, would have preferred to write a new thesis on the subject.

The experiment consisted simply of placing some pieces of fresh meat in jars which were covered with cloth. The meat spoiled in the usual way; but no maggots appeared. So accustomed are we now to meat without maggots that it is difficult to realise how astonishing was such a discovery. It was further noticed that large flies, always seen around putrefying meat, crawled over the protecting cloth and laid their eggs there—as near to the meat as they could reach. It was not hard to arrive at the truth of maggots: they grow from eggs laid by flies and are not created by the process of putrefaction.

This simple and homely experiment, unheralded in the annals of those times and still ignored by historians, was one of those forward steps of the human intelligence which mark the opening of new epochs. The concept that living things originate only from parents of their own kind and not from putrescence was a necessary foundation stone of biology. Not that Redi was at once believed: far otherwise. In his day the microscope was beginning to illumine the biological world with new and undreamed-of facts. In 1683 Anthony van Leeuwenhoek announced the existence of the minute creatures which we now call the bacteria.

They were found in a variety of materials but particularly in decaying organic substance. Putrescent meat or broth swarmed with these "animalcules," often in motion; even the yeast taken from a fermenting liquid was seen to consist of millions of obvious living cells, capable of growth and multiplication. It seemed clear that this profusion of microbia could only have come into being as a result of the decomposition of the parent material. As life ceased, the atoms of the body took on new activities, vibrated in new patterns, resuming their animation in millions of small particles instead of as members of the one great particle.

Learned men, following Redi's example, performed new experiments, as cogently planned as his to test the origin of this minute multitudinous life. Beef broth was put in a bottle and covered with hot ashes so that any living thing in it must be killed; it was corked up so that no minutest living thing could enter. Yet, after a few days, the broth became turbid in the bottle; when the vessel was opened the evil odors of putrefaction were evident and when a drop of the broth was examined with the microscope, the usual swarm of animalcules was seen.

Thus did the science of the day prove that there is in organic substance a certain "generative power," which, under appropriate conditions, cannot fail to form new life from the materials of death. Truly, it would not at a bound produce maggots, highly organized creatures; but it became manifest in the appearance of minute and elemental forms, the simplest possible living things.

Fortunately there were not lacking those who were still ready to question current conclusions and to devise new experiments to test them ever more rigorously. The broth was boiled for a long period, the flasks were sealed instead of corked. Sometimes the liquid remained unwontedly clear and when removed from the flask was as sweet-smelling and free from life as when it was fresh—an apparent contradiction of nature. Sometimes, alas! the usual signs of decomposition appeared. Bitter grew the war of conflicting ideas. Those who managed to produce sterile flasks, in which no life appeared, were accused of having tampered with the "conditions necessary to the production of life" rather than being credited with a demonstration that life is not indeed produced.

Elaborate apparatus was devised to admit air—a necessity of the "generative power"—to the broth and yet exclude any possible air-borne particles which might be alive; so originated the cotton plug, still standard in every bacteriological laboratory. In spite of every care, every precaution, there were still some flasks in which the tell-tale cloudiness appeared, in which the microscope revealed living creatures.

Until late in the nineteenth century, two hundred years after Redi's work, the question of spontaneous generation was still troubling the scientific world. Nor was it a matter of only theoretical import. Microbes were found also in living bodies afflicted by diseases; some animals which died of disease swarmed with parasitic life. Spontaneous generation accounted for these also as products of the sick activity of the diseased tissues; while the proponents of the contrary view had to hold—most unplausibly, it seemed-that the living microbes had been introduced from the outside; that in fact the microbes caused the sickness, not the sickness the microbes.

The question eventually produced the man who could settle it. Louis Pasteur, after an abortive attempt to be an artist and after having obtained a grade of "mediocre" in his school chemistry, had in a few years attained distinction by noteworthy discoveries in that science. His researches led him into the study of the fermenting juice of grapes; after he had demonstrated his mastery of its normal chemistry he was asked to search for the causes of the abnormalities that sometimes occurred, the souring and spoiling that marred the vintage.

He was not slow to see and to study the yeast cells and

bacteria that were present. He isolated them and caused them to grow in liquids of his own devising. He was the first to attain to some technical control of their life. His laboratory became filled with "cultures" of microscopic plants; a sort of garden very familiar to modern biology, but very new to his time.

As a result of his familiarity with bacteria and the like Pasteur became imbued with the idea that they did not differ essentially from other plants and animals. They resembled them in the foods which they needed and in the uses to which they put these foods. Why, then, assume an origin different from that of all other living things? Could they not come from parent microbes which because of their small size might float undetected in the air or adhere to the walls of flasks? Might they not perhaps be more resistant to killing than larger living things? Was not spontaneous generation the last medieval myth of biology?

When he reached this point in his reasoning (having meanwhile solved important practical problems for the growers of grapes and brewers of beer) and announced his intention of tackling the vexed question of spontaneous generation, his friends and admiring colleagues threw up their hands; certain of the failure to which any such attempt was predestined, they endeavored to dissuade him. Fortunately for us they did not succeed.

Within a few years that powerful and fertile imagination, building on the technique acquired in years of patient research, had devised experiments which could leave no doubt of the issue. He showed that sufficient heat would indeed kill every living thing; boiling under pressure was sometimes necessary. He proved that an organic liquid so sterilized would "keep" indefinitely provided it were not contaminated by particles from the outside world.

He showed that the flask might even be left open to the air and decomposition not occur;—if the neck were drawn into a narrow tube which was bent at a sharp angle. As air passes into such a flask any solid particles in it come in contact with and adhere to the moist walls of the crooked tube and never reach the interior. The climax of the demonstration was to shake some of the unspoiled liquid from the flask into the crook-neck; where it promptly spoiled and generated its swarm of microbes.

Nothing wrong with its "generative power"! He demonstrated that air does actually contain living microbes capable of infecting organic substance; he filtered them out of air in a cotton plug and so proved their existence and their frequency. He even made a trip to the ice fields of the high mountains to collect air which (in the absence of living bodies or other source of organic material) was free from microbes.

His researches, announced in vigorous and combative words, supported by beautiful and conclusive experiments, finally succeeded in uniting all the strands of thought that had come down from the time of Redi and weaving them into an intelligible fabric. The last recalcitrant was finally convinced, as biology began to assume its modern shape. *Omne vivum e vivo!*

From this point Pasteur went on into those studies of disease which have made his name to shine in the annals of human suffering. With his background he could not fail to believe that diseases—at least some diseases—are caused by microbes rather than the reverse. After long and sometimes bitter conflict with the entrenched medical conservatism of his day, he carried the point. The great span of modern medicine may be said to rest on two main piers: the discovery of bacteria by Leeuwenhoek in the seventeenth century and the demonstration by Pasteur in the nineteenth that they can cause disease. But that is another road, which we cannot here follow.

Omne vivum e vivo. All life, even the most lowly, is from antecedent life. The living things found in decaying matter are the progeny of other living things and the cause of the decay; not the result of it. Decay, dissolution, putrefaction—by whatever name we signify our distaste for the passing of things back into the dust—these are manifestations of the life of minute living creatures, visible to our eyes only when aided by powerful magnifiers.

Were it not for this microscopic life so abundant all about us, the earth would be littered with the desiccated corpses of everything that had ever died upon it, except as they had been removed by fire. This is the modern meaning which may be attached to the idea expressed by Carlyle. Decay is truly a form of life—but not of the life of the thing that has died. That substance is now serving another life, the life of microscopic plants.

It seems strange to many that we characterize microbes as plants. Surely they have few of the attributes of plants. Not only are they extremely small, they have no leaves or flowers or fruit or any other parts by which one might classify them; most significant of all, they have no chlorophyll.

Fig. Indian Pipe Monotropa

It is to be noted at this point that a number of quite obvious plants have no green colour. The Indian pipe of the deep woods rears soft stems of ghostly pallor from which hang nodding white flowers. The cancer-root has pretty flowers of a pale lavender on white leafless stems. The squaw flower of the western mountains is a thick and graceless flowering stalk coloured a vivid scarlet. Dodder twines its orange smother around the smartweeds. The largest of all flowers has an immense and repulsive perianth strongly suggestive of carrion, which is embellished by no green leaves.

There is no doubt that these are indeed plants; we are forced to abandon colour as a criterion for distinguishing the plant and animal realms. So also, on many counts, are mushrooms plants and toadstools and puffballs and molds and yeasts and bacteria. Heterogeneous as this assemblage is, it is held together in our minds by a common lack of chlorophyll; which has the most far-reaching effect upon the lives of its members,—and, incidentally, upon our own.

Plants with chlorophyll lead a kind of life denied to all other beings. The energy which they use is absorbed by them directly from the light which falls upon them. The green pigments are a means of trapping this energy; once transformed it finds its outlet in the creation of organic substance, food. By the decomposition of that food all the living cells of the plant (whether they have aided in its manufacture or not) are energized into that activity which we name life.

Animals take energy from foods in much the same way as do green plants; but, lacking chlorophyll, they must take the foods from some source outside their own bodies. They are dependent upon plants for the original absorption of the sun's energy, for the storage of it in organic substance.

Plants which lack chlorophyll are in exactly the same predicament. They are wholly dependent upon green plants for food. Since in their nutrition they must compete with animals, with ourselves, they assume a place of particular importance in our lives.

Some such plants, named saprophytes, use as food the remains of green plants which litter the earth and become incorporated into soil; dead and withered leaves, twigs, trunks, fruits, seeds. The cellulose of those dead cells is food, providing some organism has enzymes which will cause its digestion; starch and other common foods may linger in the dead remains.

Other nongreen plants are parasites; they send roots or some equivalent part directly into the bodies of other living plants or animals and steal the food almost at its source. Dodder pushes small pads of tissue into the supporting stems and saps their food and water; cancer-root similarly attacks the roots of other plants beneath the ground, usually some goldenrod or wild aster.

The flowering plants which lack chlorophyll are in other respects clearly related to green plants. Their flowers, their fruit and seeds, are not essentially different from those of other flowering plants. Most of the plants which share their way of life are flowerless and without vascular structure; in some respects recalling the pondscums and seaweeds.

They are the Fungi, a varied multitude of different sorts of plants: mushrooms and toadstools, puffballs, brackets, molds, slimy growths of various kinds, yeasts; some growing in leaf-mold rich in food, some on or in bread and cheese and jam and jelly and fruit; some within living tissues of man, beast, or flowering plant.

The bacteria themselves, those little demons of medical science, are often included in this group; though, because of their peculiar place in our affections, generally treated as a separate tribe. In contrast with most other plants, pleasant things which we welcome into our homes and use in our arts and industries, we usually regard the fungi with something of revulsion. They are associated with rotten wood, with damp and dark places, with spoiled and ill-smelling food or cloth or refuse.

The Fungi also include those molds which, introduced into the curd of milk, yield flavors prized by the connoisseur of cheese. The molds grow, decomposition occurs and the waste products gratify the palate of the epicure. Yeasts also are fungi; yeasts, familiar as small white squares wrapped in foil, will attack starch and sugar, producing alcohol, carbon dioxide and other substances; they have contributed largely to human welfare and not a little to human sorrow. Even if we exclude the germs of human disease, it is plain that the group is no mere plaything of the curious botanist but an object of vital concern to civilized man. Whether it is a question of making or of saving foodstuffs, fungi enter the situation and must be controlled.

Persons uninitiated into these mysteries have a fundamental difficulty in grasping the idea that these things are really plants and really alive; especially in understanding that they are of many different species, even as are the plants in other and more familiar groups. To such persons germs are just germs: a contamination, a pervasive evil. Mark Twain wrote: "For every dead body put into the ground, to glut the earth and the plant-roots and the air with disease-germs, five or fifty, or maybe a hundred, persons must die before their proper time; a dead saint enters upon a century-long career of assassination the moment the earth closes over his corpse."

He did not know that the bacteria which cause the decomposition of dead bodies are not the same as those which cause death by disease. They are everywhere present and are powerless to attack living bodies. The disease-causing organisms, the pathogens, far from causing the decomposition of their dead, are unable to live unless they find a new living body in which to multiply. Few will remain alive and dangerous beneath the ground.

Those that cause death have been called the "inefficient parasites" because they did away with their own means of livelihood. Furthermore, the pathogens themselves are different from each other. That which causes tuberculosis can never cause cholera, nor vice versa; when Bacterium tuberculosis multiplies by division, all its progeny are Bacterium tuberculosis and no others.

The bacteria form species which are just as enduring, just as distinct as species of lilies or ferns. It is true that they cannot be distinguished by visible peculiarities, colour, form and the like. But their characters are such discoverable things as idiosyncrasies of diet, limitation to particular environments (living body or non-living refuse), their production of odors and other by-products of decomposition, their connection with the symptoms of definite diseases.

Similar things may be said of the thousands of species of proper fungi. The mold that springs up on a damp piece of bread is not just an exudation from the bread, a chemical product of starch and moisture; it is a living, growing plant, which can be identified and given a Latin name as surely as Lilium regale or Rosa carolina; which grew from a spore which was formed on a parent like itself and which itself can reproduce and form new mold plants of the same type. Though they are not evident to a casual glance, the characters which distinguish these plants are as precise as those which enable us to identify roses and lilies.

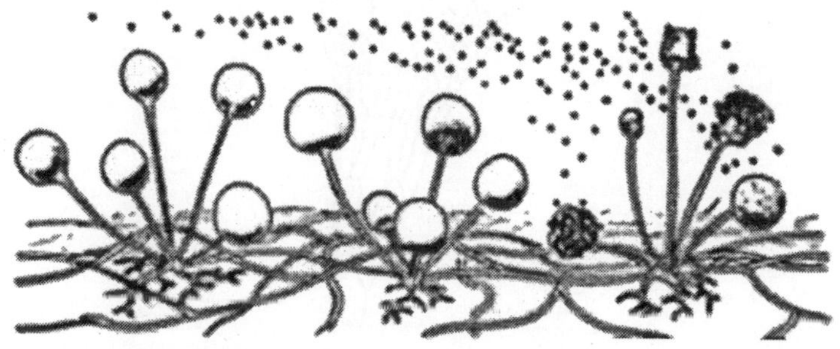

The common black mold which appears on bread or fruit is distinct in many ways from the green mold, having a different type of structure and forming its spores and gametes differently. Not only do they differ in visible features, they are physiologically distinct: neither is suitable for ripening cheese. So also the various species of yeasts and all the legion kinds of mushrooms, toadstools, puffballs and earthstars are distinct species, able to reproduce their charming or revolting characters, coming true to seed as surely as the green herbs and trees among which they grow.

A piece of fresh moist bread, left exposed to the air, quickly molds. Mold appears likewise—unsown, unsought — upon jelly and preserves, upon coffee or melon-rinds, upon damp paper or leather, upon almost any food or beverage or article manufactured from plant or animal tissues and containing some water. Before the growth of biology it was natural to assume that a mold was simply the result of dampness and air, a chemical phenomenon. Now we know that it is a plant, its delicate branched threads filled with living moving protoplasm, growing, branching, forming new substances and new parts. Mold comes only from previous mold; wherever a patch of mold is seen, a spore or a cluster of spores has settled out of the air, absorbed water, germinated.

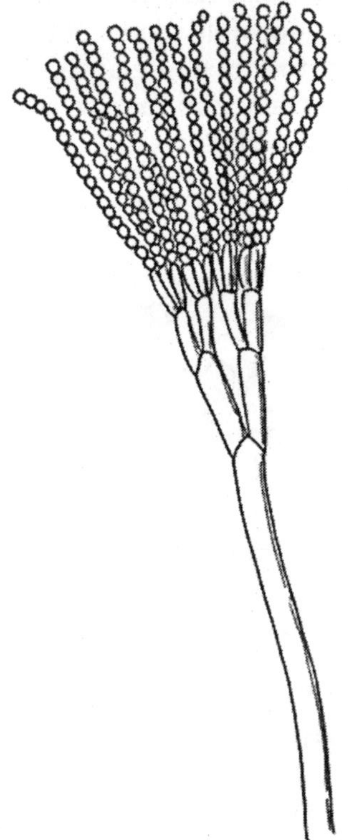

Fig. Penicillus

Part of the lack of interest which characterizes the usual attitude towards these plants is due to their apparently amorphous nature. When we really look at them, when we see that they, like peonies and daffodils, have a characteristic and indeed beautiful structure, indifference and dislike change to admiration and interest. One kind of mold, for instance, known as Rhizopus, forms its multitudes of spores in dainty globes held aloft on slender stalks.

They are liberated by the disintegration of the shell which surrounds them and drift away, leaving only the enlarged knob-like end of the stalk which upheld them. The green molds named Penicillium owe their name, which means a brush, to their spore clusters. These are formed of branching threads, the ends of which are pinched off as if by invisible scissors; the severed ends remaining attached and the process being repeated beneath them, until each branch ends in a chain of these spores.

The bodies of these plants are formed, as has been said, of narrow threads, branched and interlaced to form a web. The filaments of Rhizopus are like those of certain Algae in lacking cross walls; they are transparent tubes filled with streaming protoplasm, in which drift thousands of small nuclei. The plant is coenocytic, non-cellular; though the relations between nuclei and surrounding cytoplasm are doubtless comparable to those in other plants.

Penicillium has threads regularly cut by transverse partitions into a series of cells placed end to end—again like some Algae; these cells, however, may contain pairs of nuclei instead of single nuclei. In general it may be said that the molds have various types of organization comparable to all those found in Algae, with additional complexities which they themselves have invented; a puzzling, highly developed group of plants, inconformable to the general rules of structure established for green plants.

Molds also reproduce through the formation and union of gametes; a single web, more versatile than the average flowering plant, may form several distinct kinds of propagative bodies at the same time. When two webs of the black mold Rhizopus are intermingled, so that threads of each are in contact, a curious union occurs. At the point of contact the threads sprout small side-branches, which become greatly distended; their swelling pushes apart the filaments from which they sprang. Each of these now club-shaped branches becomes divided by a cross wall into a terminal part and a basal part; the two terminal parts are in contact.

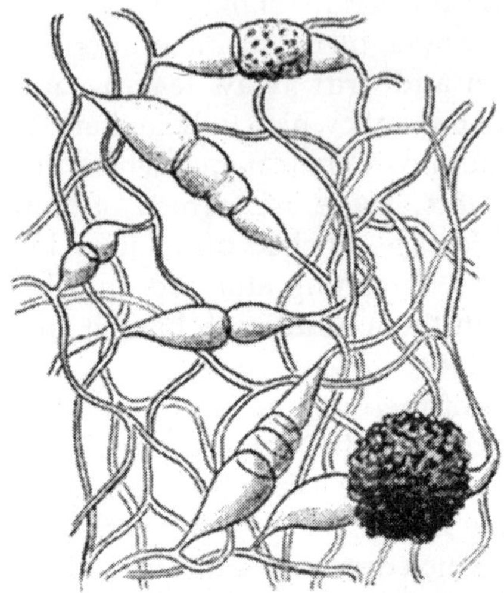

Presently the wall between these two disappears, their separate outlines are lost and they coalesce into one spherical body. This zygospore forms a thick black rough coat and within this shell the protoplasm may remain alive for long periods, dormant; eventually germinating to form a new mold filament.

In this type of reproduction there is little difference between the bodies which unite. In some relatives of Rhizopus, indeed, they may be to all appearances identical and even arise from threads of the same mold plant. In other fungi a distinction is made between two types of gametes. Such molds as Pythium, dreaded of gardeners for the "damping-off" which it causes, form large egg-like gametes and smaller, narrower bodies which grow towards them and discharge their contents into them. In the great group of Fungi of which Penicillium is a member, the history of reproduction is exceedingly variable and complex.

Cells may come in contact, nuclei may migrate from one cell to another in a seemingly random fashion. Often every cell of a fungus web contains a pair of nuclei, one descended from each of two original parents. In some final cell the two nuclei unite, completing the reproductive union initiated many cell-generations before.

Around the zygote thus formed may grow various complex structures; small spherical hard bodies beautifully ornamented with spines or hooks, inside which are delicate membranous sacs which contain spores; or goblets of various sizes whose inner surface is lined with the like spore-containers. To detail the manifold structural inventions of this group would necessitate a textbook; merely a glimpse may here be given into this field of arduous and fascinating study.

Fig. The Visible Part of a Mushroom is a Reproductive Body, a Sporophore

Of all Fungi the mushrooms and their kin are the most obvious, the most familiar, to most persons the most representative. We see them everywhere in moist woods, along paths, near old wooden steps and porches, even in the fields; springing up with an uncanny promptness after a spell of rainy weather. On a lawn a tree was cut down, but the roots were left in the soil. For many years thereafter, whenever rain fell for a week or so at any time during the warmer parts of the year, clusters of brown toadstools would appear, flourish for a brief time and wither away.

The fungus was perennial in the woody remnants of the tree, feeding on the organic substances, emerging into a conspicuous reproductive development whenever water was abundant enough for rapid enlargement.

The solution to the mystery of the rapid appearance of toadstools and the like is that they dwell continually in the soil, perhaps for long periods giving no visible sign of their presence. The body of such a fungus is rarely seen by the amateur; difficult to isolate even for the expert; for it consists of a web of delicate threads something like that of a mold, which penetrates the soil and is entangled with it, entering the narrowest interstices as do the root hairs of a flowering plant.

The filaments are formed of chains of cells end to end like those of a pondscum. They branch and rebranch and often become massed in larger tougher strands, almost root-like, which may more easily be detected around the base of a mushroom by a little patient delving. This is the actively feeding part of the fungus; it can flourish only in soil which contains organic matter, remains of plants that have died—for having no chlorophyll and no light it can make no food, absorb no energy from the outer world.

The visible part of a mushroom is a reproductive body, a sporophore. When certain conditions of temperature and of moisture prevail, a small mass of threads grows from a region on the underground web and organizes itself into a compact round button. This pushes up rapidly through the soil, breaking the surface, even pushing rocks aside or-most wonderfully—cracking paving-stones with its soft body; such is the osmotic force of its thousands of minute cells. Within the button a cylinder of gills is formed. The cap flattens, raises its margins, carrying the gills up on its under surface, standing like an umbrella on its narrow stalk; from the lower end of which its margin has been torn.

Of the many curious features of this pale urgent being not

the least curious is its internal structure. If cross sections are made no pattern of differentiated types of cells appears. If small fragments are dissected with the points of needles, it becomes evident that the whole body is made up of threads. It is simply a portion of the underground web, an outgrowth which kept its filaments more or less parallel and closely associated. We touch here on one of the greatest unsolved riddles of biology: how such an assemblage of separate threads, each growing by virtue of foods and energies contained within itself, can conspire to realise a body of a certain form.

A flight of pigeons turns and manoeuvers in the air without losing the outline of the whole; so do these undifferentiated threads, having no particular form of their own, together create a body of definite shape and colour and recognizable as belonging to a certain species. Botany has as yet no inkling of a mechanism by which this may be accomplished.

The mode of organization is entirely different from anything in the vascular plants, the ferns and conifers and flowering plants; so is the form that is achieved; it may be that the mechanism which brings form out of masses of living units is the same through the plant kingdom.

Mushrooms and toadstools are of many colours, delicate

tints of lavender or pink, creamy yellow, or gaudy orange or red; never chlorophyllous. The edible species are grown in cool damp cellars, often in abandoned mine-workings and caves. Many edible kinds are never cultivated but are recognized and gathered in nature by the expert connoisseur. Such habits are unfortunately dangerous, for many species are poisonous, some deadly. This is generally known and it is perhaps because of this knowledge that men speak of mushrooms as "unhealthy-looking" and of their beautiful colours as repulsive. But it is not so generally known which are the poisonous kinds.

Fig. Boletus

One of the commonest questions directed at botanists by the rest of the world is: "What is the difference between mushrooms and toadstools?" The implication is that the toadstools are the undesirable kinds and that the botanist at least can discriminate between them and the edible or harmless mushrooms. With regret the botanist has to answer that there is no difference; that there are no such groups as toadstools and mushrooms.

Edible and poisonous species may be members of the same family of fungi, even of the same genus, only to be recognized by the botanist who has devoted some time to them. Their edibility does not correspond with easily visible characters by which they may be classified and identified. Most of the victims of fungus poisoning thought they knew the mushrooms, had some supposed criteria of innocence.

Unfortunately there is no sure sign of the poison within them, no signature writ upon them by a beneficent Creator. The common superstitions, the timehonored household tests that make use of silver spoons, the methods of preparation that involve acids,—all are fallacious; each has killed its hundreds and thousands.

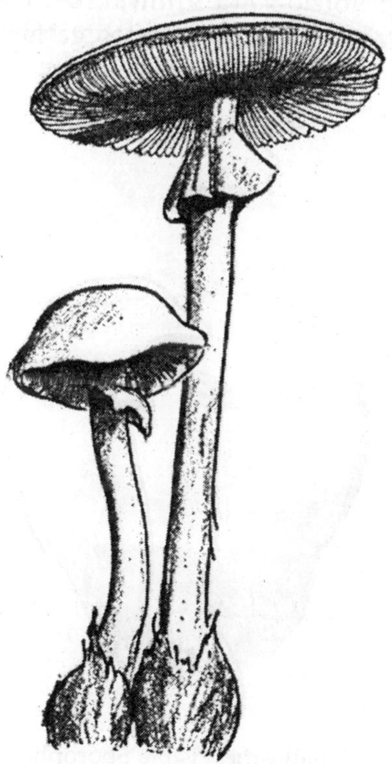

Fig. Destroying Angel

The poison distilled by these enemies of animal life is of many kinds. In some fungi it may not cause death. One who eats one of these may undergo violent pains, nausea, vomiting; but if medical help is quickly available, the sufferer may live to avoid toadstools in the future. But a few species contain a more terrible venom, as cruel, as deadly as that of a cobra. For some hours after it has been swallowed it gives no sign; then it strikes, destroying living cells beyond repair.

Death is almost the only escape. The deadly Amanita, the Angel of Death, haunts many a pleasant grove. It is of an appropriate deathly pallor, its gills even being white; its slender stalk rises from a bulblike base, the "poison cup." Because experience is necessary in the recognition of this and all other fungi, because a slight mistake can be so catastrophic-avoid all fungi you do not know, trust only a person of considerable botanical experience to gather them for your table.

Fig.. Puffballs: the Visible Sporophores of a Subterranean Fungus Web

A curious incident of some poisonous mushrooms is that it is possible for at least some men to develop a tolerance to them. This happens in parts of Siberia; the natives eat quantities of a species which is decidedly poisonous to other men. It is true that they also are poisoned; but the symptoms are not vomiting and agony, nausea and coma, but —a desired intoxication! In which connection it is worth noting that the liquors which we value for a similar purpose are sufficiently poisonous to destroy smaller animals and plants, or to preserve their remains.

Other fungi, more or less closely related to the gill fungi, take various and curious forms. Puffballs are the visible sporophores of a subterranean fungus web. At maturity they are composed almost entirely of spores, contained in the dried thin outer shell. The number of spores is astronomical. It has been calculated that a puffball three inches across may contain thirty billion spores—and puffballs have been recorded three feet in diameter and weighing 47 pounds.

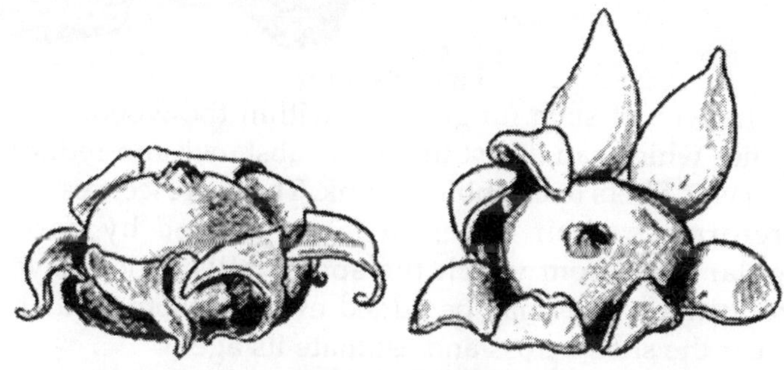

Fig. Earth Star

Most of the incredibly abundant life thus lavished upon the earth is wasted; but it is small wonder that Aristotle took as a clear example of the origin of living matter from dead substance the appearance of fungi in a pile of manure.

Still more curious are the less known members of the group: bird's-nest fungi, earthstars and others. Within a cup-like nest of fungous tissue lie three or four small lightcoloured eggs, each something like a small puffball.

These may be violently discharged from the nest and may adhere to leaves or twigs which they encounter in their trajectory; from their points of attachment they let themselves down on silken threads as a caterpillar or spider does and sway in the wind, discharging their contribution of spores.

Earthstars are puffballs whose outer layers split into several segments which curl back to form the star; curl so far back that they raise the body of the sporophore above the ground. Other groups of related fungi are soft, slimy, less definitely formed: one rejoices in the name of "Jew's ear." Still others are the brackets of shelves which project from the trunks and branches of trees.

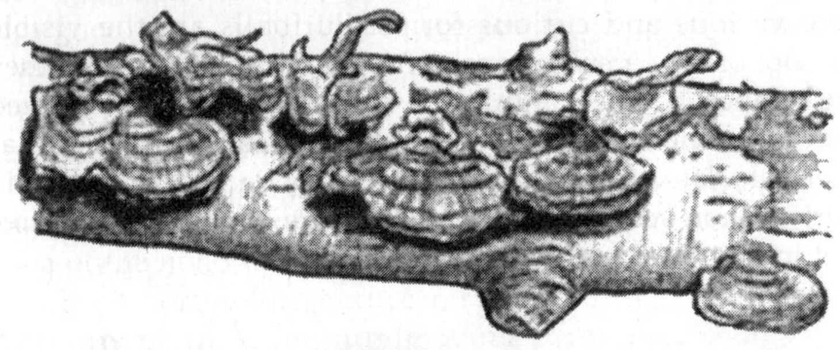

Fig. Polypore

The web of such fungi grows within the wood, forming enzymes which can digest the solid substance and reduce the firm elastic fibers to crumbling punk. The excrescences are the sporeformers; their lower surfaces pierced by pores or irregular slits, from which the spores fall. A new layer of reproductive tissue may be added every season; so that one may cut the shelf across and estimate its age.

Spores, spores and yet more spores; the air is full of them.

It seems sometimes that the air must—if these things be true—be less gaseous than solid; how is it possible to see through such a cloud of minute floating bodies? Spores of fungi, of bacteria, of ferns and mosses, pollen of pine and birch and ragweed, in numbers like the sands of the sea and the stars of the heavens. But all minute. Ordinary milk may contain 300,000 microbes in a milliliter—about 300,000,000 in a quart; yet all of these, if collected together, would be smaller than a pinhead.

The air is well populated, but is not, as one writer has put it, "no better than a stirabout of minute shapes."

There are not so many species of fungi as of flowering plants. Nor do they form a dominant part of any landscape.

Ubiquitous and successful as they are, they are less impressive than the green plants of gardens and fields. But they differ among themselves even more than do the different orders of flower-bearers: they are the most varied of plants. It is mainly convenience that has marshalled them into one group, not any rational knowledge of their relationships, of their descent.

They are small, difficult to study; their soft bodies have left few fossils in the earth's history. In our ignorance we assemble them on the basis of their common lack of chlorophyll, their dependence upon outside sources of organic matter (with their lack of vascular tissue); so we find we have made a group contain such diverse plants as a mold and a puffball, the slimy orange excrescence from a rotten twig and the beautifully formed earthstar or mushroom.

The consequences of the physiological defect of the Fungi have already been sketched. Not only do they feed like animals on the living bodies or dead remains of other organisms; they feed in a remarkable variety of ways. Scarcely any organic matter is immune to their ravages, though a particular species may be quite limited in its diet.

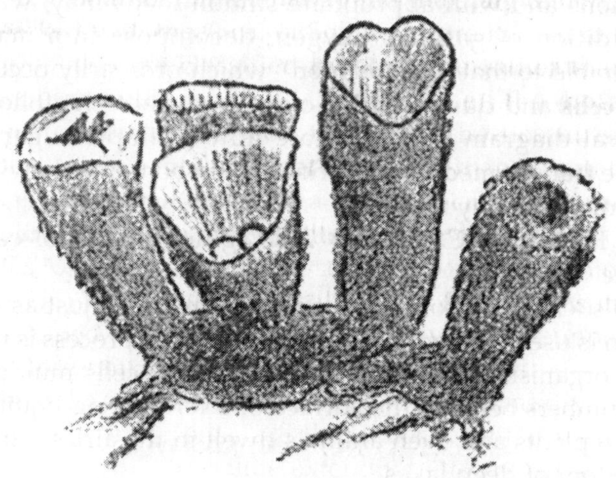

Fig. Bird's Nest

Part of the secret of their universality is the enzymes, the digestive juices, which they can form. They are almost unique in the entire vast living world in being able to digest cellulose, even wood, converting it into soluble sugars which will enter their bodies and serve as food. Thus can tender threads, composed mainly of water enclosed by a thin and flexible membrane, insinuate themselves into the toughest fabric and leave only ruin behind them. Many fungi are individual also in the chemistry of their respiratory process.

All living things obtain energy through the decomposition of organic foods into other substances. Carbon dioxide and water are common among the wastes. Some fungi, some bacteria, habitually decompose their sugars into alcohols and acids; the fermentation of the juice of grapes is a sign of the energyrelease in thousands of minute rapidly multiplying yeast cells, the bubbling gas and the desired intoxicant are the wastes of that process.

The lactic acid of sour milk has a similar origin. Many of these organisms carry on that process which impresses the layman as a contradiction in terms: oxidation without the use of free oxygen. To explain this would necessitate too great a detour into chemistry. Suffice it to say that under appropriate conditions and with appropriate stimuli foods may, without the addition of gaseous oxygen, decompose in a manner comparable to that "combustion" which ordinarily occurs in living cells and during which oxygen is used. The following chemical diagram will illustrate how a simple sugar may become transformed into alcohol and carbon dioxide without the addition of any material:

$C_6H_{12}O_6 \rightarrow 2C_2H_5OH + 2CO_2$ glucose alcohol carbon dioxide

In such a transformation energy is released just as when oxygen is used (though not so much) and the process is useful to the organism in the same way. Thus do cells multiply in vast numbers beneath the surface of a fermenting liquid and thus do plants and even animals dwell in the airless slime at the bottom of deep lakes.

Because of the easy dispersal of their innumerable spores, because of their ability to consume almost any organic material, the fungi come into frequent conflict with us. The crimes of the group, however, are not limited to the spoiling of our bread and jam, the molding of leather and the rotting of timbers.

Some of them—like the bacteria which are often included with them—feed on *living* food. They insinuate their narrow threads into the living tissues of other plants, of the plants of our gardens, fields and forests; penetrating between the cells, absorbing food from them, even killing them. These are parasites; their presence is usually signalized by the symptoms of a disease.

Disease is a familiar enough concept; a concept, however, somewhat clouded in the average mind. A disease is "something wrong"; but we commonly exclude the results of accidents, such as broken limbs or bruises. In truth our ideas of disease (like most of our ideas) are historical; they are heavily tinged with the beliefs of our medieval ancestors. These good people—clearer-minded sometimes than us though inferior in information—did not attempt to distinguish between "mental" and "physical," any more than the modern psychologist.

Melancholy was a disease as clearly characterized as jaundice or plague or scurvy and equally to be explained by some sort of disharmony or disturbance in the proper functioning of the parts of the body. Ludicrous as were many of their explanations, ineffective as were their treatments, their definitions were often correct.

When the modern scientist attempts to define disease, he arrives at a similar concept: disease is an abnormality in functioning. The distinguishing feature of the modern view is that we can catalogue diseases not only according to symptoms but according to a variety of causes, causes which are not the products of a speculative imagination but are known from good evidence. Scurvy is due to a deficiency of a vitamin. Plague is caused by a bacterium.

Measles is the result of infection by a filterable virus (whether a living parasite or not it is hard to say). Certain conditions which may result in "insanity" are initiated by certain genes inherited from the preceding generation. A "sick headache" may be caused by an overdose of fat or alcohol or the fumes of an automobile. All these are diseases in the sense defined. Or rather they are the symptoms of diseases, the fundamental abnormal condition sometimes being concealed. The headache may be caused by a variety of conditions, some of which may include the attacks of parasitic microbes.

It comes as a shock to many to learn that plants also are subject to diseases and in the same sense that we are. This harks back again to the traditional concept that plants are not truly alive, that biology is the study of animals. So a farmer, noticing rough rust-coloured spots on his wheat and rightly correlating their appearance with reduction in vigour and the yield of his acres, refers to the condition as "rust" and further noticing that its onset is most marked in periods of wet weather, concludes wrongly that the phenomenon is indeed analogous to the rusting of iron. Actually this is a disease due to the invasion of the wheat by a parasitic fungus, a plant of great complexity and menace; the study of which indicates methods by which the losses which it causes may be combated.

Like our own troubles, plant diseases are the symptoms of many different kinds of abnormalities. Various discolourations of leaves, certain swellings and malformations, are due to deficiencies of certain nutrients in the soil or to the presence of poisonous substances and can often be quickly cured by the simple addition to the substratum of the material that is lacking. Other diseases are caused by vagaries of temperature, wind, or light, by insects and mites which nibble and suck at the green parts or at the roots underground, even by overdoses of the sprays or fumes with which we seek to control the ravages of such pests. Perhaps the most serious of plant troubles are caused by parasitic fungi.

If the wheat escapes rust, it may be afflicted by smut, or be variously blighted, blasted, spotted, mildewed, or scabbed. Every kind of crop has its own particular diseases; the more popular the harvest, the greater the number of subtle enemies that may deprive us of it. Apples and appletrees are subject to the attacks of rusts, scabs, spots, leafspots, cankers, galls and so on to an incredible number. And—marvelous to relate—

each of these diseases is the sign of the presence of some particular species of fungus; it is often true that a species of fungus affects only certain species of flowering plants, perhaps only one. The rust-fungus that invades an apple is powerless against wheat and though black spots are common on a variety of leaves the dreaded black spot of rose leaves is caused by a fungus which infects no other plant.

If a new crop is developed by human husbandry, new diseases appear to vex it with a promptitude that stimulates moral reflections on the power of evil. This is not our world unless we fight for it and the fight is unremitting. The familiar potato is a fairly recent addition to our economy.

Though it was discovered and brought to Europe from South America in the sixteenth century, it was grown first as a botanical curiosity; it did not, apparently, occur to botanists or to others to imitate the brown-skinned natives who fed on it in its native mountains.

After one hundred and fifty years of such indifference, it began to be used as a crop and its use and popularity spread readily, particularly in poor soils inhabited by wretched peasantries.

Fig. Disease of Potato: Late Blight

Thousands of Germans found in the potato a way to ward

off starvation and pay their landlords and armies and the Irish made it their own and gave it their name. As the fields were extended, diseases of the potato appeared; most serious was the *late blight*, which attacked the full-grown plants, killed them with deadly suddenness and spread like wildfire from plant to plant, from field to field. Every year it was worse.

The result was widespread human misery, even starvation, in the potato lands. So does an obscure microscopic fungus, known only by a polysyllabic Latin-and-Greek name, leave its mark on human history. Though the textbooks of history may name it not, it was a factor, perhaps the decisive factor, in the waves of emigration to the United States. The transcontinental railroads were thrown rapidly across the country by "cheap Irish labour"; the newcomers also found police forces to be staffed, politics to be played; the form and colour of our municipal life would have been different if there had been no epidemic of potato blight in Ireland in 1845.

To such problems as confronted the peasant potatogrower a hundred years ago the modern botanist finds simple solutions. The late blight is caused by a fungus, a moldlike plant, which grows within the leaves, between the cells, taking from them their food and pouring out in exchange its deadly poisons; Phytophthora, plant-eater.

It sends some of its branches forth into the air by way of the stomata of its victim; at the tips of these threads are formed spores, which float away in every breeze. If a spore encounters the moist surface of a leaf of another potato plant, it grows into a new mold plant, which penetrates the living leaf and kills it. Such discoveries suggest the remedy. It is impossible to cure a diseased plant, for fungus and host-plant are so intimately mingled that what hurts one will damage the other also—both being living protoplasm.

But it is comparatively easy to prevent healthy plants from taking the disease. All that is necessary is to spray or dust the leaves at intervals with a substance poisonous to the fungus and harmless to the leaves. The fungus spore at the moment of germination is a delicate plant, lacking protective covering, exposing its rapidly growing young protoplasmic threads to such a method of attack; the leaf-surface over which it grows is covered with cutin, through which water and watersoluble poisons will not readily pass. A convenient and effective poison is found in the well known Bordeaux mixture, which contains various salts of copper poisonous to protoplasm. Such prophylactic measures are now a commonplace of agriculture; late blight causes no more famines.

Other diseases of other crops have yielded to similar methods of control. The botanist identifies the cause, the fungus, perhaps growing it in glass dishes or tubes on nutrient substances; he charts its life history and chronicles the method and progress of its infection of other plants. Such systematic studies usually reveal the weak spot in the cycle of growth and reproduction; the naked heel of Achilles. The method of control is not always by using a poisonous spray or dust. Oats are heated gently and accurately to a temperature which has been found to kill the incipient smutfungus within them without damaging the embryos.

To ward off certain diseases it is necessary to sterilize the soil: to so treat it that all living things in it are destroyed, before your seedlings are set out in it.

By such a procedure docs the grower raise crops of seedlings under glass in defiance of the dread damping-off, which otherwise can lay them low within a few hours. Sprays are the commonest preventives, useful against insects as well as fungus-spores. They are indeed so common that they in turn have created a fresh problem.

They are poisonous, or they would be of no use; they are unfortunately poisonous not only to fungi but to other organisms which seek to devour plants; including man. They commonly contain lead or arsenic, often both; deadly even in small quantities. Various regulations have sought to enforce washing the fruit or leaves served on our tables, but the efficacy of the practice is open to question; it is extremely difficult to remove traces of the poisons from fresh fruits and vegetables. This is a serious problem which still awaits research.

Of the innumerable diseases of plants one of the most

curious and baffling is that first mentioned here: the rust of wheat. It got its name from the appearance on the stalks and leaves of streaks of a rusty red colour, slightly raised and roughened. If these marks were numerous, the plant was stunted and the head of grain below normal in yield or even entirely lacking. Many millions of dollars have been lost to wheat farmers through the effects of rust.

For a time it was thought that there were two kinds, the red rust just described and "black rust," which usually appeared later. It is now clear that both of these are symptoms of the same disease. The rusty streaks are masses of spores, small red spores raised on slender stalks and pushed out through the skin of the wheat.

These red spores float through the air in their hundreds of thousands and infect new victims; germinating more easily, of course, if the weather is wet and moisture adheres to the stalks and leaves of the wheat. Later the same fungus web which has formed and discharged these spores and which continues to live and grow within the tissues of the wheat plant, forms spores of another shape and colour, black spores, which are likewise discharged through the broken epidermis of the host. This is the black rust.

Now we encounter the most surprising feature in the disastrous cycle: these black spores *cannot infect wheat*. They first remain dormant for a long period which lasts perhaps through the winter. At the end of this dormancy they germinate, forming a small thread on which very minute spores are borne. These latter spores also can do no damage to wheat; they can develop to maturity only if they find haven on the leaves of another flowering plant, quite unrelated to wheat: the common barberry.

On the leaves of this host appear round swollen spots, within which the web of fungus growth may be discovered. The fungus does not, indeed, do much damage to the barberry. Such a state of affairs seems to lack all reason. Why should one sort of spore of a fungus infect wheat and nothing else, while another spore of the same fungus, with the same heredity, can develop only if it finds a barberry bush? How could such a relationship have originated? Wheat and barberry are quite unrelated and probably never grew in the same field until man brought them together comparatively late in his history. Upon what mysterious chemistry do the "preferences" of the spores depend? And do our own whimsical likes and dislikes, our desires and revulsions, have a similar basis?

The barberry completes the cycle; for in the same unpredictable way the spores formed by the fungus in the barberry will not germinate on the plant parasitized by their parent, but only if they are carried to wheat. These spores are formed in small cups which project usually from the lower surface of the infected spot; their development entails some very complex unions of gamete-like bodies produced by the fungus. If they succeed in finding wheat, the spores thus created can germinate, send their threads into the tissues of the new host and initiate a new outbreak of rust in the wheatfield.

It occurred to scientists in the United States Department of Agriculture that if there were no barberry it would be a good thing; the black spores and their progeny would find nothing they could infect, the red spores would be killed by winter; the next spring there would be no diseased barberry from which spores could infect wheat.

The principle is just that which is involved in the extermination of rats in cities ravaged by bubonic plague, or of mosquitoes in countries afflicted by malaria. The fight against barberry has been long and costly. Barberry bushes are hard to find in the woods and extremely hard to eradicate. And unfortunately it has become clear that much of the rust-fungus can survive the winter without the intervention of the barberry, at least in the more southern states where winter wheat is grown.

Index

T

U

V

W